The First Americans

A H I S T O R Y O F U S

BOOK ONE The First Americans

BOOK TWO Making Thirteen Colonies

BOOK THREE From Colonies to Country

BOOK FOUR The New Nation

BOOK FIVE Liberty for All?

BOOK SIX War, Terrible War

BOOK SEVEN Reconstruction and Reform

BOOK EIGHT An Age of Extremes

BOOK NINE War, Peace, and All That Jazz

BOOK TEN All the People

Oxford University Press

In these books you will find explorers, farmers, cowboys, heroes, villains, inventors, presidents, poets, pirates, artists, slaves, teachers, revolutionaries, priests, musicians—the girls and boys, men and women, who all became Americans....

OXFORD
A HISTORY OF
US

BOOK ONE

The First Americans

Joy Hakim

Oxford University Press
New York

Oxford University Press
Oxford   New York   Toronto
Delhi   Bombay   Calcutta   Madras   Karachi
Kuala Lumpur   Singapore   Hong Kong   Tokyo
Nairobi   Dar es Salaam   Cape Town
Melbourne   Auckland   Madrid
and associated companies in
Berlin   Ibadan

Copyright © 1993 by Joy Hakim

Designer: Mervyn E. Clay
Maps copyright © 1993 by Wendy Frost and Elspeth Leacock
Produced by American Historical Publications

Published by Oxford University Press, Inc.
200 Madison Avenue, New York, New York 10016
Oxford is a registered trademark of Oxford University Press

Library of Congress Cataloging-in-Publication Data
Hakim, Joy.
The first Americans / Joy Hakim.
p.   cm.—(A history of US: 1)
Includes bibliographical references and index.
Summary: Presents the history of the Native Americans from earliest times through the arrival of the first Europeans.
ISBN 0-19-507745-8 (lib. ed.)
ISBN 0-19-507746-6 (paperback ed.)
1. Indians of North America—History—Juvenile literature.
2. America—Discovery and exploration—Juvenile literature.
(1. United States—History. 2. Indians of North America—History.)
I. Title.II. Series: Hakim, Joy.History of US: 1.
E178.3.H221993 vol. 1
[E77.4]92-42507
CIP
AC

3 5 7 9 8 6 4 2
Printed in the United States of America
on acid-free paper

"This Newly Created World" is taken from the Winnebago Medicine Rite as recorded in Paul Radin, *The Road of Life and Death: A Ritual Drama of the American Indians* (p. 254), Bollingen Series V, copyright © 1945 by Princeton University Press; copyright © renewed 1972 by Princeton University Press. Reprinted by permission of Princeton University Press.

This book is for Sam, who always believed in these books and in me.

It is for all those who read the manuscript and supported and encouraged the idea of a children's storybook history.

It is for Byron Hollinshead, who made it possible.

And it is for you, too, dear reader.

Most of the pictures in this book were drawn in the time the book describes. Many of them were done by the artists John White, Jacques LeMoyne, and Theodore de Bry. The cover illustration was done in the 16th century by a companion of Hernando Cortés. It shows Aztec warriors confronting the conquistador Pedro de Alvarez and his men.

Picture Credits

This Newly Created World
(a Winnebago Indian poem)

*Pleasant it looked,
this newly created world.*

*Along the entire length and breadth
of the earth, our grandmother,
extended the green reflection
of her covering
and the escaping odors
were pleasant to inhale.*

The Cause of America is in a great Measure the Cause of all Mankind.
—TOM PAINE, 18TH-CENTURY POLITICAL LEADER

America is not like a blanket—one piece of unbroken cloth, the same color, the same texture, the same size. America is more like a quilt— many pieces, many colors, many sizes, all woven and held together by a common thread.
—JESSE JACKSON, 20TH-CENTURY POLITICAL LEADER

All other nations have come into being among people whose families had lived for time out of mind on the same land where they were born. Englishmen are English, Frenchmen are French, Chinese are Chinese, while their governments come and go; their national states can be torn apart and remade without losing their nationhood. But Americans are a nation born of an idea; not the place, but the idea, created the United States Government.
—THEODORE H. WHITE,
20TH-CENTURY AMERICAN HISTORIAN

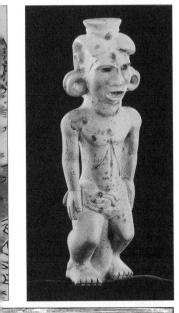

Contents

1 History? Why? — 11

2 Away with Time — 14

3 In the Beginning — 16

4 How the First Americans Became Indians — 19

FEATURE: SOME THOUGHTS ON DINOSAURS AND THE EARTH — 22

5 Put On Your Earmuffs — 24

6 Cliff Dwellers and Others — 27

7 The Show-offs — 31

8 Taking a Tour — 34

9 Plains Indians Are Not Plain At All — 40

10 Mound for Mound, Those Are Heavy Hills — 43

FEATURE: DIGGING UP DIRT ON THE MOUND BUILDERS — 47

11 Indians of the Eastern Forests — 48

12 People of the Long House — 52

13 Let's Turn North — 56

FEATURE: HOW DO YOU KNOW WHAT YOU KNOW? — 60

14 The Power of the Press — 61

15 A Boy Named Christopher Has a Dream — 64

16 A New Land Is "Discovered" — 67

FEATURE: FROM COLUMBUS'S PEN — 71

17 The Next Voyage — 72

18 Stowaways: Worms and a Dog — 77

19 Sailing Around the World — 80

20 What's in a Name? — 85

21 About Beliefs and Ideas — 87

FEATURE: STORIES OF HOW THE WORLD BEGAN — 90

22 New Spain 91

FEATURE: TENOCHTITLAN 96

23 Ponce de León, Pizarro, and Spanish Colonies 97

24 Gloom, Doom, and a Bit of Cheer 100

25 North of New Spain 102

26 Being a Conquistador with Coronado 106

27 Conquistadors: California to Florida 110

28 A Place Called Santa Fe 115

29 Las Casas Cares 117

30 The Big Picture 122

31 From Spain to England to France 124

32 France in America: Pirates and Adventurers 126

33 Rain, Ambush, and Murder 130

34 New France 133

FEATURE: TRAVELING BY CANOE AND PORTAGE 136

35 Elizabeth and Friends 138

36 Utopia in America 140

37 Lost: A Colony 143

FEATURE: FROM JOHN WHITE'S LOG 147

38 An Armada Is a Fleet of Ships 148

39 The End: Keep Reading 150

CHRONOLOGY OF EVENTS 152

MORE BOOKS TO READ 154

INDEX 156

A NOTE FROM THE AUTHOR 160

1 History? Why?

The ancient Greeks believed in nine goddesses called *muses,* who inspired the arts—things like dancing, music, and poetry. This is Clio, the muse of history, in the Capitol, Washington, D.C.

What's the point of studying history? Who cares what happened long ago? After all, aren't the people in history books dead?

Those are good questions. They bother a lot of people. In fact, they bother some people so much they never study history. That's too bad, because those people miss out on something very important: their own story.

History is the story of US. It tells who we are and where we have been. Sometimes it is so surprising it jolts your mind. Here are a few answers to the questions about studying history:

History is full of stories—true stories—the best ever. Those stories have real heroes and real villains. When you read history, you are reading about real-life adventures.

History is a mystery. No one knows what happened in the past—at least we don't know the whole story. We weren't there. Have you ever put a jigsaw puzzle together? That's what learning history is like. You gather pieces of information and try to discover how they fit. Suddenly, when you have enough pieces in place, you begin to see the big picture. That's exciting, and so is studying history, because new pieces of the puzzle keep fitting in.

When we read about *the mistakes people made in the past*, we can try not to make them ourselves. Nations and people who don't study history sometimes repeat mistakes.

History, after all, is the memory of a nation.
John F. Kennedy
35TH PRESIDENT OF THE UNITED STATES

History is more or less bunk.
(AND, AT ANOTHER TIME):
The farther you look back, the farther you can see ahead.
—*Henry Ford*
FOUNDER, FORD MOTOR COMPANY

The first law for the historian is that he shall never dare utter an untruth. The second is that he shall suppress nothing that is true.
—*Cicero*
ANCIENT ROMAN HISTORIAN

It is one thing to write like a poet, and another thing to write like a historian. The poet can tell or sing of things not as they were but as they ought to have been, whereas the historian must describe them, not as they ought to have been, but as they were, without exaggerating or hiing the truth in any way.
—*Miguel de Cervantes*
FROM HIS NOVEL *DON QUIXOTE*. CERVANTES DIED IN 1616, THE SAME YEAR AS SHAKESPEARE

A man without history is like a tree without roots.
—*Marcus Garvey*
FOUNDER IN 1917 OF UNIVERSAL NEGRO IMPROVEMENT ASSOCIATION

What experience and history teach is this—that people and governments never have learned anything from history, or acted on principles deduced from it.
—*G.W.F. Hegel*
A 19TH-CENTURY GERMAN PHILOSOPHER

History is especially important for Americans. In many nations—Japan or Sweden, for instance—most citizens share a common background. They have a similar look. They may worship in the same church. That isn't true of us. Some us were once Chinese, or Italian, or Turkish, or Ethiopian. We don't look alike. Sometimes we don't think alike. But as Americans we do share something. It is our history. We Americans share a common heritage. If you are an American, then the Indians, the Vikings, the Pilgrims, and the slaves are all your ancestors. You will want to know their stories.

Learning about our country's history will make you understand *what it means to be an American.* And being an American is a privilege. People all over the world wish that they, too, could be American. Why? Because we are a nation that is trying to be fair to all our citizens. That is unusual.

Which brings us to our theme. It is this: *We believe the United States of America is the most remarkable nation that has ever existed. No other nation, in the history of the world, has ever provided so much freedom, so much justice, and so much opportunity to so many people.*

That is a big statement. You don't have to agree with it. Arguing with a book's theme is okay.

Some people will tell you of evil forces in the United States. They will tell of past horrors like slavery and war. They will tell of poverty and injustice today. They will be telling the truth.

We didn't say the United States is perfect. Far from it. Being fair to everyone isn't as easy as you may think. (Do you treat everyone you know equally? How about people you don't like?) The United States government has made some terrible mistakes. It is still making mistakes. But usually this nation can, and does, correct its mistakes. That is because we are a democracy: power belongs to the people, not the rulers. We are also a nation governed by law, and that is very important. No one is above the law. Everyone—the President, congressmen, congresswomen, and you—lives by the same laws.

Our top—or supreme—law is the Constitution. Even bad presidents and congresses obey the Constitution. They have to. They can be impeached—which means "brought to trial"—if they don't.

Our Constitution is part of what makes us so unusual. The Constitution of the United States was the first written constitution—in all of world history—to attempt to treat each citizen equally. It begins with the words "*We the People…*" You are part of "the People."

The Bill of Rights—which is another name for the first ten amendments to the Constitution—guarantees us rights you wouldn't want to be without. It protects our right to worship as we wish, and it gives us the right to speak out and say or write what we want. In the United

We the People

States we can criticize the government and not worry about being thrown into jail. That isn't true in some countries.

When you say the Pledge of Allegiance, do you ever think about the meaning of the words "*with liberty and justice for all*"?

Liberty is freedom, and you know what that is. But think for a minute what it might be like to live in a country where you are not free. Did you know that in some countries you are told what work you must do and where you must live?

Justice is fairness. Having the same laws for everyone is fair.

The last words in the Pledge of Allegiance are "*for all*." This is a nation that tries to offer opportunity for all. In the United States you are free to do anything that anyone else can do. You can run for president, be an artist, write books, or build houses.

The more you study history, the more you will realize that all nations are not the same. Some are better than others.

Does that seem like an unfair thing to say? Maybe, but we believe it.

We don't believe that people in one nation are better than those in another. Every nation has a mixture of good and bad people.

So why, if people are the same, are nations different?

Ideas have a lot to do with it. Nations stand on their ideas. We're lucky. The architects who designed this nation had sound ideas. They were looking for liberty, justice, and opportunity when they came here. They made sure the United States provided them.

Then they did something never done before: they created a people's government. Some men and women in other parts of the world thought that was impossible. After all, it was an untried idea. But America's citizens proved that government by the people can work. How we did that is a fascinating story.

That's the story of US—the people of the United States—the story we're about to tell. It's a story of hunters, explorers, pirates, slaves, and boys and girls who came to a strange land and made it their own. It's a story with heroes—and villains.

We're going to start that story way, way, way back in time, with the very first Americans. Read on—we have much to tell.

Where there is history children have transferred to them the advantages of old men; where history is absent, old men are as children.
—*Juan Vives*
SPANISH PHILOSOPHER, BORN 1492

History never repeats itself; at best it sometimes rhymes.
—*Mark Twain*
AUTHOR OF *THE ADVENTURES OF TOM SAWYER*

As you can see, the famous people quoted here don't all agree about the importance or even the meaning of history. What is your opinion?

2 Away with Time

Early humans made these axes by chipping at flints with a bone.

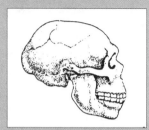
Are you ready? You're about to study the history of US—the people of the United States. You'll need to use your imagination. You can start by climbing into our sturdy time-and-space capsule. Hold on tight; you're going back to the Age of Stone and Ice. You're going to northern Asia, to a place called Mongolia.

When you get out of the capsule you might shiver a bit; it is cold here in Mongolia. But you have nice warm clothing. Because of that astounding invention—the needle made of bone—people can now wear clothes of animal hide that fit their bodies. Do you see that man and woman over there, sitting near the fire? They are your mother and father.

They are good parents, and you live happily in a fine hut built of big bones and branches and earth.

Actually, you have two homes: a winter place and a summer home. You and your family are hunters—big-game hunters—so you follow the animals to summer and winter grazing lands. You camp out when you make those journeys.

If people in the 20th century could see you, they might say you're not very different from the animals—but they're wrong. Oh, you are dirty and smelly, and you aren't wearing much—just an animal-skin

ADD FOUR FEET HERE!

FIRST POTTERY

40000 BC ◄35,000 years► BEGIN FARMING · BEGIN WEAVING

outfit. Your table manners are terrible: you rip your food apart with your hands and teeth. But you're not like the animals—you're not as fast or as strong. What makes you really different, though, is your fine brain and your hands. You use them both to make tools. You even play a musical instrument—a flute—which your father made from a piece of bone.

Most of the tools you make are of stone. (That is why this long-ago time is called the Stone Age. Later, when people discover metals and how to use them, there will be an Iron Age and a Bronze Age.) You use stones to make spearheads and axes, and you use hollowed-out stones for cooking.

A long time ago, someone discovered fire. Lightning may have hit a tree and started a forest fire. The fire cooked animals and made people warm, and everyone realized how valuable it was. Your parents make a spark of fire by striking stones together. Fire has allowed you to travel north, to this cold region, to follow the big animals—like woolly mammoths or their cousins the mastodons or polar bears.

Still, finding food isn't easy in northern Asia; sometimes you go hungry. But don't worry, things will get better. Much better. You're on your way to America. Hold on tight! It's going to be quite a trip.

No one knows when the first dogs were domesticated, but bones from 14,000 years ago show that people had dogs as companions (and perhaps as a food supply).

Mammoths were big, bigger than today's elephants. They may have used their tusks like snowplows to dig for grass in ice and snow. Those tusks could be 16 feet long.

DOGS DOMESTICATED

‹BC AD›

·COLUMBUS·

Today

15

3 In the Beginning

Ice Age bison were much larger than their descendants are today.

Watch that band of people move across the plain. They look hungry and tired. The tribe is small, just twenty people in all, and only six are men of hunting age. But they are brave and their spears are sharp, so they will keep going. They follow the tracks of a mammoth. If they can kill the mammoth—a huge, woolly elephant—they will feast for much of the winter.

The trail of the great animal leads them where no people have gone before. It leads them onto a wide, grassy earth bridge that stretches between two continents. They have come from Asia. When they cross that bridge they will be on land that someday will be called America. The trail of the mammoth leads them from Asia to a new world.

They don't realize what a big step they are taking. They don't know they are making history. All they know is that they have lost the mammoth. He has outsmarted them. But it doesn't matter; the new land is rich in animals and fish and berries. They will stay.

All that happened a long time ago, when families lived in huts and caves and the bow and arrow hadn't even been invented. It was a time when ice blankets—called *glaciers*—covered much of the northern land. We call it the Ice Age. Some of the glaciers were more than a mile high. Nothing man has built has been as tall.

If you look at a map, you will find the Bering Sea between Asia and Alaska. That is where the earth bridge used to be. It was quite a bridge. Today we call it Beringia, and it is under water. Back in the Ice Age, Beringia was a thousand miles wide. It had no trees, but was full of lakes and the kind of wild plants that drew animals. Men and women followed the animals: they settled on Beringia and lived there for generations. Not all the people were big-game hunters. Some were

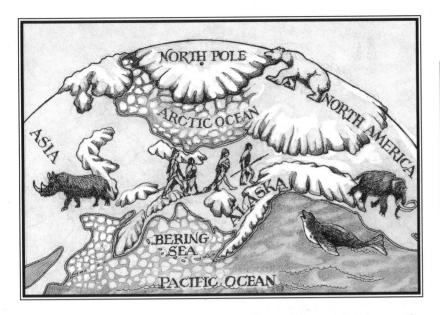

seagoers. They fished and caught small animals and lived near the beach. We think they were very good sailors. They had boats covered with animal skins, and they killed whales and explored and settled the coastline. Gradually these people—the land rovers and the seagoers—took that big step onto the new continent.

Alaska, where the hunters went, seemed like a fine place. There were seals, bison with big curved horns, birds called ptarmigans, and other good things to hunt and eat. Glaciers covered large parts of North America, but much of Alaska and Siberia was free of ice.

More hunters came with their families. At first there was plenty of food, but after a few years (maybe it was a few thousand years) the land seemed crowded. There was no longer enough game for everyone to hunt.

So each year, when the hunters saw birds fly south, they wished they could fly over the glacier. One day some people watched the animals moving south and decided to follow them. The animal herds led the people to a path, like a grassy roadway, between parts of the great glacier. It was a narrow gap, perhaps 25 miles wide with mountains of ice on either side, and it led them to the plains of North America. It was worth the trip. They found grasses and nuts and berries to eat. They found a hunter's wonderland: there were antelope, musk ox, bighorn sheep, lions, deer, moose, fox, otter, beaver, saber-toothed tigers, and bison. Some of the animals had come from Asia too, by walking over the earth bridge, or by swimming in the sea.

In America animals had grown big—bigger than any animals you have ever seen. Some beavers were as large as bears; and birds—great vulture-like teratorns—had wings that reached 15 feet from tip

About 10,000 years ago, near what is now Folsom, New Mexico, a hunter stuck a flint spear between the ribs of a bison. The skeleton lay where the bison had fallen. Slowly, it turned to stone. In 1927 it was dug up for us, the hunter's descendants, to marvel at.

The musk ox had few enemies—till humans arrived.

to tip. Lions were huge, and moose antlers measured eight feet across. (How many feet tall are you?)

You see, these animals had had two continents all to themselves. They had never had to fight the great animal enemy—man—as animals had to in the Old World. America had been an animal paradise, but that was finished. Man had arrived. America was now a hunter's heaven.

Since those first American hunters had plenty of food to eat, they started having big families, and that increased the population. After many thousands more years, people were living all over North America and down to the very tip of South America. Remember the seagoers? We think they came to America by sailing down the coast. It is possible that a few people came sailing across the oceans from distant islands or continents. Perhaps they were fishermen searching for big fish; perhaps they were lost at sea. No one is sure about that. We do know that the American continents seemed a good place for people to live—especially as the climate was changing.

It was getting warmer—slowly. Up north, the big glaciers—which were like seas of frozen water—melted and froze and melted and froze. When they melted, the oceans got bigger and flooded some of the land. The earth bridge disappeared under the sea; then, when the waters froze again, the earth bridge reappeared. The pathway through the glacier closed up and then reopened. Finally, about 10,000 years ago, things settled down much as they are now. Water covered Beringia. It got very cold in the Bering Sea. When all that happened, the animals and people who had come across to the new land were stuck. They couldn't go back to Asia, and no one could join them. America was cut off from the rest of the world. There would be no new people for a long, long time.

The camel originated in America and traveled across the Bering land bridge to other continents. The llama (shown here), found in South America, is a modern descendant of the ancient American camel.

4 How the First Americans Became Indians

First Americans grew tomatoes long before they were brought to Italy to become an ingredient of spaghetti sauce.

Those first Americans, the people who had come from Asia, were on their own in the New World. Thousands of years would pass before Christopher Columbus finally arrived. He came across the Atlantic Ocean from a distant continent. Columbus called the First Americans "Indians" because he thought he was in the Indies—way over by China. That was a big mistake. But it would be a long time before anyone realized how wrong Columbus was. So the name "Indian" stuck. It's what we use most often today.

Some people say "Native Americans" instead of Indians, although the word "native" is confusing. It has two meanings. Anyone who is born in a country is a native of that country, so many of us are native Americans. "Native" also means to have an origin, or beginning, in a country. As far as we know, no people is native to America. Our ancestors all came from somewhere else.

Still, the people who came over the Bering Strait were here long before anyone else. So you can see why it makes sense to call them Native Americans. Indians, First Americans, or Native Americans— they are all good names. (Most Indians call themselves simply "the people" in their various languages.)

If we go back in time 10,000 years, those First Americans, who now were spread out over the two American continents, were beginning to do some remarkable things.

Most of them continued to hunt—they were good at that. They knew how to make animals stampede into deep ditches or watery bogs, where they could be easily speared.

When it came to spearheads, theirs were the best. They made them

Some historians think that by the time Columbus got here, there may have been as many as 75 million Indians living in South and North America. That is a lot of people—almost a third of the population of the United States today.

The atlatl threw darts faster and more accurately than people could throw them by themselves.

A *stampede* is a wild rush of animals.

19

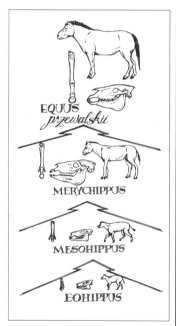

Eohippus, the horse's ancestor, lived in forests and had sharp teeth for pulling leaves off trees. It had four toes, for running in mud. As time went on, Eohippus grew bigger and moved to drier, grassier lands. Its toes fused into a hoof; its teeth got larger and flatter, better for eating grass.

of flint, a hard stone they chipped at until it was sharp and deadly.

They invented a dart thrower: a wooden handle with a hooked tip that worked like a missile launcher. It was called an *atlatl*. The hunter would throw the atlatl as a ballplayer throws a pitch. That meant he no longer had to creep up and stab or choke his prey. The atlatl probably saved a lot of human lives; it didn't do much for the animals.

In fact, so many animals got killed that some of them became extinct. (That means they all died out.) But, in most cases, their disappearance wasn't because of hunters.

The mammoths, for instance, were just too big. After the end of the Ice Age, when the climate warmed up, the mammoths couldn't adjust. They drank huge amounts of water each day; a family of thirsty mammoths could drain a pond. Scientists think they couldn't find enough water to drink or enough grass to eat.

More than one hundred species of animals (SPEE-shiz—it is the scientific word for groups of plants and animals) became extinct between 6,000 and 10,000 years ago. No one knows why all of them died. It is one of history's big puzzles.

One reason we think the hunters weren't especially to blame was that there was so much food on the land. People could make choices: to catch fish or dig for clams and oysters or gather nuts and berries and roots. And some tribes did just that. They became gatherers.

Others became farmers—among the best in the world. They took wild plants and bred them, and they developed corn, potatoes, sweet potatoes, and squash. They learned to make chocolate from the cocoa plant. They found corn kernels that popped when heated. They discovered plants and herbs that could heal sickness. They grew tobacco and peppers and tomatoes. None of these plants was known in other parts of the world.

Indian basketweavers wove baskets so tight that they could hold liquids and so handsome that people in later generations put them in museums. Potters learned to make sculptured figures and useful pots and bowls. Weavers designed colorful rugs.

Native American thinkers created mathematically precise calendars. Goldsmiths made some of the most beautiful jewelry the world has ever seen. Indians invented the hammock, the canoe, snowshoes, and a game called lacrosse. They learned to gather rubber from rubber plants; they made rubber balls and played ball games. They built pyramids and temples and cities.

But no single tribe of Indians did all those things. Native Americans developed different life-styles and different languages, just as Europeans and Africans and Asians did. It all depended on who their leaders were and where they lived.

Chasse générale du Chevreuil.

One thing Indians never did was to make good use of the wheel. It would have made their lives easier if they had. In Mexico the Indians put wheels on their children's toys, but they never made wheeled wagons for themselves. Perhaps that was because they didn't have horses to pull them.

When people came from Spain to America (at the end of the 15th century), they brought horses and mules and oxen. In the 16th century, horses completely changed Indian life, just as the automobile and airplane changed life in 20th-century America. (Can you imagine hunting buffalo on foot? Now jump on horseback and see the difference it makes.)

When the horse came to America, it was returning home. A tiny horse ancestor had lived in America in Ice Age times. Some of those ancient, dog-sized horses had trotted across Beringia to Asia. In Asia they grew large and galloped on—to Europe and Africa. Those that stayed in America became extinct.

These are pictures made centuries ago by European explorers in America. The pictures show a few of the many ways that some Indian peoples hunted and trapped animals. Check the top right-hand corner. What do you think of the contest between the deer and the hunters? How would you like to be that deer?

21

Some Thoughts on Dinosaurs and the Earth

The supercontinent Pangea, 200 million years ago, before the land split up.

Have you been wondering about dinosaurs? Did you know they once dwelled on the American continents—perhaps right where you live? That was long, long ago—about 200 million years ago, way before people and other mammals existed.

It was a time when North America, South America, Africa, Europe, Asia, Greenland, Australia, and Antarctic all made up a huge supercontinent. (Scientists have named that supercontinent Pangea, a Greek word that means "all lands.")

Big and little landmasses had moved together to form Pangea. Then they began moving apart. A space opened up between parts of Pangea. That kind of space is called a *rift*. The rift got very wide, filled with water, and became the ocean we call the Atlantic.

Does this sound like science fiction? Moving continents? Is that really true?

Yes, it's true. Continents do move. North America is moving right now—right under your feet. Don't run outside to watch it — progress is slow: this continent is moving at about the same rate as your fingernails grow. If you happen to have some time to spare—maybe a million years or, better yet, 50 or 100 million years—you may see real changes. People who study the earth—scientists called *geologists*—say that southern California could end up attached to Alaska someday. Some geologists say that California may split and form a land of its own.

Weird? Not really. Geology is surprising. If you read about it, you may soon be fascinated.

For instance: did you know the center of the earth is called the *core* and is made of blisteringly hot liquid rock? That liquid rock is called *magma*. Circling the earth's core is another layer called the *mantle*. It is made of rock, too. The mantle's rock is gushy and hot—but not as hot as the core. The mantle's magma is less liquid, more dense.

The third and outermost earth layer is the one we live on—the *crust*. It is a skin that stretches around the globe. When geologists talk of the crust, they usually describe it as a thin layer. But thin to one person is fat to another. In most places, that "thin" crust is 60 miles thick.

Do you have the three layers—core, mantle, and crust—in mind? Now, focus on the crust.

The crust is not smooth. It is wrinkled, cracked, and broken up into *plates*—seven giant plates and some small ones. Those plates are like rafts floating on the liquid mantle. They move, they crash into one another, they slide on top of each other. Sometimes they just thrash about in strange ways. (Some plates are under the sea; some hold the continents.)

Look at a map of Asia. Do you see India? India is a *subcontinent*. It was once a separate continent, far south, below the equator. Then it moved north. (It moved quickly in geological time: it took only 30 million years.) When the Indian plate hit the Eurasian plate, it didn't stop. It rammed Asia. That ramming action pushed, lifted, wrinkled the earth, and formed the Himalayan Mountains.

The Appalachian Mountains in North America were made in a similar manner. They were lifted out of the earth when landmasses crashed into each other as Pangea was forming. Those ancient Appalachians were huge when they were new—450 million years ago. Glaciers, wind, and rain have worn them down. Mountains are made in other ways, too. Look at a map of the United States. The center of this country happens to rest on especially thick crust. In dinosaur days it supported an inland sea. On either side of that crust are thin areas. Now start pushing the continent west, which is the direction it is still traveling. What happens to skin when you push it? Some skin stretches smooth, but some areas wrinkle. That is what happens to the earth. The earth's "wrinkles" are mountains. The Rockies are layers of thin, wrinkled crust thrust up from a former ocean bottom. Our central plains sit on smooth, thick earth crust.

Remember, the crust is floating on a sea of molten rock. When a crack—or fault—appears in the crust, some of the heat and energy of the magma can escape, and that may start things moving. When that happens, you have an earthquake along the fault. Earthquakes don't take millions of years to move the earth. They can move it in seconds.

The earth also has hot spots, which are like chimneys for interior heat and energy. Sometimes magma explodes out of the crust at a hot spot. That's what a volcano is all about. Magma—now called *lava*—pours out of erupting volcanoes.

The Hawaiian Islands came into being when a series of undersea volcanoes erupted. The Hawaiians are the world's tallest mountains, if you measure them from their base on the ocean floor.

But we started this discussion with dinosaurs. Do you know why they died out? Well, neither does anyone else. It happened suddenly, about 65 million years ago. Some say a climate change killed the plants they ate. Some say a giant meteorite fell to earth and created an environmental change— intense cold or heat or rain or drought.

No one is sure why the dinosaurs disappeared. If you can figure it out, you'll become famous.

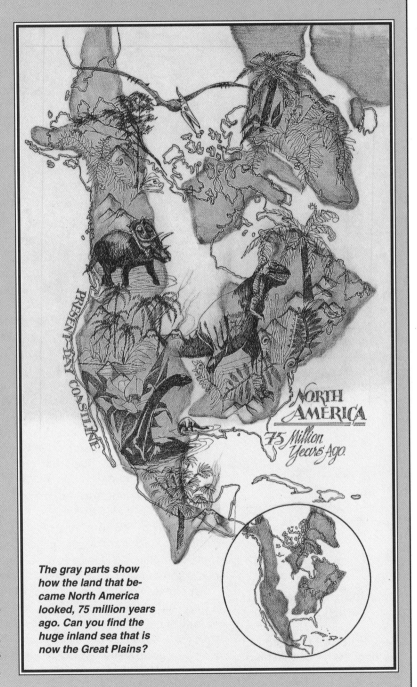

PRESENT-DAY COASTLINE

NORTH AMERICA
75 Million Years Ago.

The gray parts show how the land that became North America looked, 75 million years ago. Can you find the huge inland sea that is now the Great Plains?

5 Put On Your Earmuffs

Snowshoes were the best way to get around in soft snow.

Beringia was gone. Drowned. The land bridge was beneath the Bering Sea. America was now cut off. But a few people *did* get to America after water covered the land. Some may have come in boats, but most didn't. If they didn't come by boat and they couldn't fly and the land was gone, how did they travel? And who were they?

They were Eskimos, and they came by sled over the polar ice. The Eskimos are thought to have been the very last of the ancient Asians to arrive on the American continent. They stayed in the north country and never followed the others down to the warmer climates. Maybe they liked life in the far north. Maybe they were used to it. Maybe they were attacked when they went south.

Eskimos are not Indians but are a separate people with a language like no other. Today, most Eskimos lead modern lives. But some Eskimos live—right now—more or less as they have for thousands of years: hunting and fishing near the North Pole and across the arctic and subarctic regions in Alaska, Canada, Greenland, and Siberia. Those in Alaska are American citizens. You don't have to get into a time-and-space vehicle to learn about Eskimos. You can travel by airplane and dogsled to visit them.

By the way, the name Eskimo, like the name Indian, got attached to a people for the wrong reasons. *Eskimo* means "eater of raw meat" in Indian languages of the Algonquian (al-GAHN-kwien) family. It was those Algonquian languages that most English people first heard— and that is how the word *Eskimo* got added to the English language. The name Eskimos use to describe themselves is *Inuit* (IN-yooit), which means "real people."

Many Inuit live by fishing and hunting. They cook some of their meat and eat some raw, for the vitamins and minerals.

When the Inuit travel in winter, hunting for the fish and marine animals they eat, they live in *igluviga*, ice-block houses built on top of the iced-over sea.

It just happens that Inuits cook some of their meat and eat some of it raw. Raw meat is important in a region too cold to grow vegetables. Raw animal organs provide needed vitamins and minerals.

If ever you travel to Eskimo land, you can try some raw meat. Inuits are known for their good nature and hospitality. Perhaps that helps them survive in their cold, harsh environment. In the arctic region the temperature stays below freezing for about ten months a year. Sometimes it is -50°F. In winter it is not only cold, it is dark much of the time. In summer the sun stays bright all through the day and night!

If you go to northern Alaska, your warmest clothes will probably not be warm enough. Inuit clothes will be better. They are made of two animal skins, back to back, with fur inside and out. Under that, some Inuits wear a bird-skin shirt with feathers next to their skin. Inuit women make the garments lightweight and watertight.

Inuits who live in coastal villages hunt seals, walrus, and whales. Others, who live inland, are nomads and must move to survive. They hunt reindeer (also called *caribou*) and the shaggy musk ox. Those who hunt by the traditional methods use harpoons and clubs. They fish through holes cut in the ice. They live on land called *tundra* that has no trees and stays partly frozen all year round. Some people call their land the frozen desert. It is a desert made of ice. Winds blowing across the ice make the cold seem even colder. Snow falls only rarely.

The flap of this woman's deerskin coat is for sitting on. The coat is decorated with pewter (PYOO-tur) spoon bowls in front and a fringe of colored beads on the collar.

∗ The long arctic days and nights are caused by the earth's rotation. You can look up the details in a good geography book.

25

An Inuit carved these snow goggles out of walrus ivory.

The words **_tundra_** and **_taiga_** came into English from Russian. Why do you think that happened? (Look at a physical map of Russia and Siberia.)

Taiga, Taiga

Some Inuits live in the subarctic regions—a land with forests of pine, spruce, and aspen. That land, called taiga, can be even colder than the treeless tundra. In the subarctic, trees stop the fierce wind and leave the snow powdery; animal tracks can be seen in the snow and hunters follow them. In summer, there are berries and other plants to eat. In warm weather the snow melts and big, biting mosquitoes appear. But only about 100 days are without frost. The season is too short to farm. People of the taiga use snowshoes to get around. Because they live among trees, they can keep warm with fires.

Most Inuits live in pit houses made of double stone walls covered with earth. You must crawl through a tunnel to get inside the house. The tunnel helps keep the cold air outside. Smokeless oil lamps warm the house and are used for cooking. There are no fires because there are no trees and therefore no wood. Oil comes from melted animal blubber. (What is blubber? It is animal fat.)

When they travel to hunt and fish, Inuits live in igloos (the word *igluviga*, in the picture caption on the previous page, is the Inuit word that "igloo" comes from). An igloo is a domed house built of hard snow bricks. Light comes inside through ice windows. An Inuit can sometimes build an igloo in an hour. (Try building a dome of bricks. It isn't easy. You need to understand some engineering principles.)

Igloos may have two or three rooms connected by snow-brick tunnels. Sometimes Eskimos build their igloos together with a big central room for dancing, singing, and talking. During long winter nights people often tell stories, study the stars, and carve and decorate animal bones and walrus tusks with fish and animal shapes. In summer Inuits live in tents made from animal skins. How do they travel from place to place? On dogsleds. An Inuit family's dogs are its prized possessions.

Inuit boats, called *kayaks* (KY-yacks), are made of oiled animal skins stretched on frames of driftwood. A kayak is narrow, with a hole in the top just big enough for one person to slip inside. Its skin covering pulls tight around the waist of the paddler. Kayaks are sleek and fast and built so that if they tip over they can be turned back up again easily. The watertight Eskimo clothes keep the wearer dry and warm in icy seas. Large, open Inuit boats, called *umiaks* (OO-me-acks), which are often 40 feet long, are used for whaling. Umiaks, too, are made of skin and driftwood.

As we mentioned, you can fly to northern Canada or Alaska and visit Inuits, but if you want to know about the traditional ways of life of most other Native Americans, you will have to climb back into the time capsule.

There are many ways to hunt seals. Waiting patiently by a breathing hole is one way.

6 Cliff Dwellers and Others

An Anasazi artist painted a pronghorn antelope on this bowl about 1,000 years ago.

The Anasazi are just one of the peoples of the ancient Southwest. The Hohokam, Hakataya, and Mogollon are others. We think the Anasazi's modern descendants are Pueblo peoples like the Tiwa, Zuñi (ZOO-nyi), and Hopi.

Go ahead. Climb in the capsule and set the dial to the year 1250. You are going to the Southwest of what will someday be the United States: to the place where four states—New Mexico, Arizona, Colorado, and Utah—meet. You are going to Mesa Verde, Colorado.

That's where the amazing Anasazi (an-nuh-SAH-zee) Indians live. But they don't call themselves Anasazi—that's the modern name for them. We have no idea what they call themselves. *Anasazi* is a Navaho word. The Navaho Indians are nomads, wanderers who sometimes raid and steal from the peaceful Anasazi farmers. The word Anasazi, in Navaho, means "ancient enemies." Like Eskimo and Indian, Anasazi is another of those wrong names that history sometimes attaches to people. The Anasazi are resourceful people. This should be an interesting visit.

Imagine you're an Indian boy, not quite a year old. Your mother is carrying you to the fields. You are strapped to a board on her back. Hold still or this could be a real cliff-hanger. (Being strapped to that board will make the base of your head flat. The Anasazi see that as a sign of beauty.)

You've just left your home, which is part of a 200-room apartment house built on a natural stone shelf on the side of a steep mountain. To get to the cornfields, where she will work, your mother has to go the rest of the way up the mountain—straight up—by pulling herself from one toehold to another. Don't look down: the canyon floor is 700 feet below. (Imagine looking down from the roof of a 70-story building—you would be looking down about 700 feet.)

Usually your mother stays home, takes care of you, cooks, cleans

Name Calling

The Navaho nomads, and other tribes called Apache, are said to have come from western Canada. Someday they will settle in villages and become farmers, shepherds, and silversmiths. A lot of the names we use for Indians now are based on names other people gave them (often rude ones). For instance, the Navahos were called the *apache* of *navahu*, "enemies of plowed fields." Do you think the people we call *Apache* called themselves that?

27

The cliff has plenty of clay for making jars and pots.

the house, and makes pots out of snake-shaped coils of clay. She uses the sharp leaf of the yucca plant to paint black designs on the white pottery. But at harvest everyone is needed to help gather the corn, squash, and beans that grow on the flat top of the table-like mountain.

This year the skies have been generous with rain. A man-made reservoir is filled with water. The corn is heavy. Soon you will see your first harvest dances: creatures wearing painted masks will jump and dance to the pounding of drums. You will tremble and maybe scream when you see the eagles, the wolves, and the ferocious giants. You will smile at the dancing corn maidens and butterflies and laugh at the painted clowns. The ceremony is meant to thank the gods for the harvest and to prepare for the hunts and harvests to come. (Only when you are grown will you learn that these are people of the community. Then you will take part in the dance, too.)

While people in faraway Europe are building cathedrals and going on crusades, you are living in a stone castle tucked under a mountain roof. You Anasazi are like swallows nesting in the hollow of a hill; you are protected from heavy snows and from human enemies, too. But the stones are damp and the apartments cramped. As soon as you are grown, you will begin to feel the pains and aches of arthritis; you will die before you are 40. (Centuries from now, scientists will study your bones and learn these things.)

Crusades are religious expeditions and wars. In the 11th, 12th, and 13th centuries, armies of European Christians went on crusades to the Holy Land, Jerusalem. Jerusalem had been conquered by Muslims, and the Christians wished to recapture it.

Still, it is a splendid home. The mountain site faces south and catches the sun's rays. In winter, with a roaring fire on the town's flat plaza and fires in each house, you are warm even on snowy days.

These are new buildings you live in. Your parents were born in a village on top of the table mountain, a village with fine houses and lookout towers. Your people have lived in this region for hundreds of years. Why has everyone moved to the side of the mountain? (Perhaps you can leave a clue; future generations will want to know the answer to that question.)

It must have been very hard to build these rooms where they are. Building materials had to be carried up, or down, the steep side of the mountain. Your apartment house is a marvel of architecture. It has walls built of heavy stones, held together with thick clay. There are towers too, and many kivas. The kivas are round rooms dug into the ground. Men gather inside the kivas to make laws, to discuss problems, to hold religious ceremonies, and perhaps just to have a good time. Sometimes there are games on the roof of the kiva.

The Anasazi weren't the only people who died at what we consider an early age. The average lifespan in England or Spain in the 13th century wasn't much longer. People often died from common diseases, or malnutrition, or just by getting caught in a war.

Before you leave your teens, a girl will sit in front of your door for four days grinding corn. If she grinds well, and pleases you and your parents, you will marry her. As a wedding gift you will weave yucca fibers into a pair of sandals and put them on her feet. Your parents will give the two of you a blanket of turkey feathers.

But right now your father is waiting for you on top of the mountain. Like most of his friends, he is a farmer. This is a sharing community, and deeply religious. There is no freedom to hold different religious views here—but there is tradition and order and harmony. The priests are the most important people in your town, but day-to-day affairs are run by a council of town leaders. Your father is a member of the all-male council. Of course, you are too young to understand all that, but you are not too young to love music. Your father has a flute, made from a reed, and he plays it sweetly. He is also a good ballplayer,

Ball games at Mesa Verde are of the minor-league variety. You can find major-league play at a pueblo in Arizona where the ball court is as big as a football field and is lined with 20-foot-high walls.

Every day the Anasazi farmers climbed to the flat top of the mesa to work in their fields. The round chambers of the pueblo are kivas.

This Zuñi scarecrow and the crows' dead bodies make a gruesome team for keeping birds off the planted fields.

and in a few years you will be able to play ballgames with him.

Your name is Swift Deer. Your parents hope you will become a fast runner. There are no horses in this land, no animals to ride, so runners are important. But you may surprise your parents and become a mighty hunter and chase and kill rabbits, pronghorn, deer, and elk. Maybe you will become a trader and travel to far-off Mexico, or bring shells home from the western sea. Perhaps you will be a runner after all and take messages to Chaco Canyon. Your Anasazi cousins have built a 2,000-room complex of apartments and official buildings on flat land there. Long avenues, like wheel spokes, lead out to smaller villages.

Is it imagination and artistry that have caused your people to build a great city in a cliff? Or is it a need to feel safe? Can you imagine anyone trying to attack a cliff?

But your cliff home does not keep you safe from drought. In the year 1276, when you are a grown man and a father yourself, terrible times begin. For the next twenty-four years there will be little or no rain. Your people did not practice conservation during the good years. Many trees were cut and now there is not enough wood for fires. The land has been overplanted, which means that the nutrients in the earth that feed the crops have been used up; harvests begin to shrink. You and your friends go hungry. By the year 1300, when you are in your grave, the cliff will be empty of people.

Your children and grandchildren will go south to the great river that will be called Rio Grande. They will build small villages and live in homes made of clay that is strengthened with sticks and brush.

These clay-plastered villages are called *pueblos* (PWEB-lowz). The sun-dried clay mud is called *adobe* (uh-DOE-bee). A flat-topped mountain is called a *mesa* (MAY-suh). Mesa Verde (VAIR-day) means "green table mountain." Those are all Spanish words. (Keep reading this book, and you'll find out how the Indians met the Spaniards.)

To see the pueblo in your mind, imagine rectangular rooms made of clay. On top of one is a smaller room set back a bit, then another on top of that, and another. Picture stairs. The roof of one house is the foundation and front yard of the house above. There are no doors on the ground floor of these houses. To enter you must climb a ladder and go through a hole in the roof. When enemies approach, the ladders can be raised. Put a lot of the houses together, like town houses, and you get the idea of a pueblo village. It is an efficient way to build.

The Southwest is a hard place to grow crops; it is too hot and too dry. Yet Pueblo Indians, more than any other Native Americans, depend on farming. Those who live near the river are able to irrigate their fields by digging ditches that bring water from the river. Usually they can survive dry times, but in the desert region it is never easy.

7 The Show-offs

One view of an eagle: at the top are two profiles, facing each other, of its head and beak.

In case you forgot, you're still in that time-and-space capsule, but you're not a baby anymore. You're ten years old and able to work the controls yourself. So get going; we want to head northwest, to the very edge of the land, to the region that will be the states of Washington and Oregon. The time? We were in the 13th century; let's try the 14th century for this visit.

Life is easy for the Indians here in the Northwest, near the great ocean. They are affluent (AF-flew-ent—it means wealthy) Americans. For them the world is bountiful: the rivers hold salmon and sturgeon; the ocean is full of seals, whales, fish, and shellfish; the woods are swarming with game animals. And there are berries and nuts and wild roots to be gathered. They are not farmers. They don't need to farm.

These Americans go to sea in giant canoes; some are 60 feet long. (How long is your bedroom? Your schoolroom?) Using stone tools and fire, Indians of the Northwest Coast cut down gigantic fir trees and hollow out the logs to make their boats. The trees tower 200 feet and are 10 feet across at the base. There are so many of them, so close together, with a tangle of undergrowth, that it is sometimes hard for hunters to get through the forest. Tall as these trees are, they are not as big as the redwoods that grow in a vast forest to the south (in the land that will be California).

These Native Americans carve animal and human figures on tall fir poles, called totem poles. The poles are painted and are symbols of a family's power and rank. The Indians' totem poles are colorful but rough; finer poles will be carved after the Europeans come and bring metal knives.

Because food and wood are so easy to gather, the Northwest Indians

The names of some of the Northwest tribes are: Kwakiutl, Tsimshian, Tlingit, Nootka, Chinook, Makah, Haida, Okanagon, Spokane, Quinault, Kalapuya, Kalispel, Shuswap.

In Europe important families also have symbols of power. They design crests. The crests are picture symbols—or totems—that are put on castles and armor.

31

have much leisure time. Their lives are full of playacting, dancing, and singing. In times of celebration, relatives and friends come from far villages. Drums are held near the fire until their animal-skin tops can be stretched across their birch frames. Then these people of the coastal forests gather in a circle and dance and sing of the fish and animals they will hunt. They also sing of their ancestors, and of their fears and hopes, and they pray to the animals for forgiveness and for good luck in the hunt. Sometimes they have wrestling contests. Often they wrestle just for fun.

A tightly woven basket hat kept its wearer nicely dry on the rainy coast of the Northwest.

A warrior of the Tlingit people in cedar-slat armor ready for battle. The head carved on his helmet helps indicate his tribe and rank.

Sometimes the best wrestler gets to marry a special girl.

Many Indians elsewhere in North America live in communities where almost everything is shared—sometimes even leadership. That is not true here. These Indians care about wealth, property, and prestige (press-TEEJ—it means importance and reputation). They value private property, and they pass their property on to their children and grandchildren. They own slaves and sometimes go to war with other Indians just to capture slaves. People are not treated as equals in this society. They are divided into ranks, or classes. There are slaves, commoners, and nobles. In times of strife, many of the men become warriors and wear wooden helmets and wood slat armor.

These Indians of the Northwest Coast like to pile up their goods and show off. They have good taste. They weave handsome blankets, make beautiful baskets, carve fancy wooden bowls, and fashion spoons of decorated animal horn. Their dress-up clothes are gorgeous. Their houses are spectacular. Sometimes several families live together in a large house built of wooden planks with carved and painted walls and posts.

Captain Cook

When the English sailor Captain Cook comes here in 1778, he will be horrified by how slaves are treated by Indians of the Kwakiutl and Nootka clans. Some people may find the captain's reaction surprising because, in his lifetime, English ships sail from England to Africa to America with captured Africans who are sold into slavery. Yet many people, in England and America, don't approve of slavery. Do you think that is true of some Northwest Indians?

They take pride in other possessions, too—in furs, copper shields, and fancy hats.

There is something about these Indians that no one in future times will quite understand. It has to do with parties. They are big partygivers. Sometimes they spend years planning fabulous parties called *potlatches*. There is much feasting at a potlatch. The party may go on for days and days. Then the host gives away his finest possessions; sometimes he gives away everything he has.

Have you heard of people spending years planning a wedding? Do you know of people who spend more money on a party than they can really afford? Perhaps these Indians are just vain and

The word *potlatch* is from a Nootka word, *patshatl*—"giving." Perhaps members of the Haida clans represented by these raven and frog totems gave away things as beautiful and costly as this Chilkat blanket.

boastful, like other people in other times and places. Perhaps there is an important reason for the potlatch.

Some of the guests at the potlatch will plan their own parties and try to make them bigger and give away more things. The bigger the party and the more that is given—or sometimes even thrown away—the more prestige the giver has.

Was the potlatch a way to gain power? Or to show off? Or something else? No one really knows.

8 Taking a Tour

A potter from Casas Grandes in California fashioned this jar about 700 years ago.

Usually when we say *tribe,* we mean a small community. A *people* is a larger group all speaking the same language.

In this chapter we're going to move in time and space. You'll be busy working the controls of the capsule. So get ready. We need to take a spin around this land to see the big picture. Remember, we were at Mesa Verde in the 13th century and in the Pacific Northwest in the 14th, so let's try the 15th century for this trip across the continent.

It will help if you are good at learning languages, because the Indians living in North America speak at least 250 different languages. That's a big problem. The tribes often don't get along well because they don't understand each other. (Do you think that happens with nations today?) Don't look for a typical Indian; there is no such thing. Indians live differently in different environments.

Take us up high over the continent. Are you looking down? Does what you see look a little like the maps in your schoolbooks?

No. You're right, it is much prettier.

Start over there, on the West Coast, where we just left a tribe having a potlatch. Then you can head south, along the foggy coast. Notice the sandy beaches and the rocky cliffs that drop right down into the ocean. Watch out! Don't fly too low or those redwood trees will scrape the bottom of this space vehicle. Some of them are 300 feet tall. If you look carefully, you will see tiny orange-and-blue specks at the feet of those giants. The specks are tiger lilies and iris, and they thrive in the moist soil under the towering trees.

Do you see that plank canoe? We count 14 oarsmen on each side. They are skimming across the water as fast as some birds fly.

The land carriage

We are still on a course south and heading into the sunshine of the land that will someday be called California. It is a land of plenty. Look below us. Most of those California Indians you see are easygoing, gentle people. Some tribes have lived here for thousands of years. The women weave colorful designs into watertight baskets and then use the baskets to gather acorns. The acorns are ground into meal that makes nourishing bread. There are many different peoples in this region speaking many different languages. Mostly, the tribes live peacefully with each other. The trade routes between the villages are well worn. Because food is plentiful, these people have much time for games, music, storytelling, and religious festivals.

Look at the steam coming from that building! Is it a health club? Inside, men sit and sweat. Soon they will dive into a cool stream to bathe. The steam huts are something like the kivas we saw in the land of the Anasazi: they are both social halls and spiritual centers.

Now put the capsule on a course east. That flat land, spreading back from the coast, will someday make fine farmland. Even from here, way up, we can see wildflowers—fields of orange poppies, purple thistles, and flowering mustard the color of canary birds. Check out the mountains; you may have to zoom close to see how rugged they are. Wait until the pioneers try to cross those snow-coated California mountains with their covered wagons! They will need ropes to pull the wagons up one side of the mountain and ropes to lower them down the other side. They are formidable, those mountain peaks—steep, rocky, and hard to climb.

The desert, which we are now approaching, isn't easy to cross either. Notice the cactus: its red and gold flowers match the reds and golds of the desert sun. Indians eat the giant cactus and often rest in its shade, but if anyone makes the mistake of leaning against a cactus—ouch!—he'll never do it again.

See that violet carpet? Those blossoms on the desert floor are lavender plants. The pink flowers are verbena.

But if we spend all our time looking at wildflowers, we'll never get anywhere. So let's do some zooming: over incredible wind-carved stones that stand gaunt and lonely, like giant statues on the high plateau the Indians call Utah; over nature's cathedrals—rock fortresses with turrets and towers all made by winds and water; and over

Snowy Range

The Spanish were the first Europeans to arrive in California (this book tells you about that, too). So it's not surprising that a lot of place names around California are Spanish. Those snowy mountains are called the *Sierra Nevada*. In Spanish, *sierra* means "range"; what do you think *nevada* means? (The state of Nevada was named after these mountains—but, unlike them, the state is mostly desert, hot and dry.)

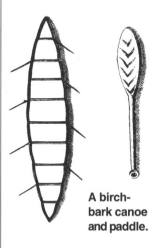

A birchbark canoe and paddle.

These great boats of elmwood are sturdy enough to be paddled standing up. Smaller, lighter canoes are made of bark.

canyons—like mountains in reverse—as much as a mile deep.

You can slow down a bit to see the soaring, wrinkled, beautiful, and young Rocky Mountains. (They are only 65 million years old; they appeared about the same time the last dinosaurs died out.) Pull up, high, so we get a look at the two continents. Can you see that the Rockies are part of a backbone that runs from Canada all the way down to the tip of South America? Those Rocky Mountains are the rooftop of this continent. They are easy to cross because of a flat area—called the Wyoming Basin—that cuts through them. The mountains are teeming with life: with tiny shrews, soaring falcons, fat grizzly bears, lordly elk, hardworking beavers, and nimble mountain goats. And, talking about wildflowers! Well, we had better not. We need to keep moving. So on we go, over the windy high plains—full of gophers and buffalo and noisy magpies—until we reach the flat lowland in the middle of the country.

The Indians were ingenious when it came to catching buffalo. Here, the buffalo follow the fences the Indians have built. At the narrow point is an earth ramp that slopes up steeply and then drops down into a corral (the square on the right). The buffalo stampede up the ramp and fall over the edge into the corral. Too bad for the buffalo; they are trapped—and doomed.

Once all this was a shallow inland sea, but that was long, long, long ago. Slowly the sea disappeared, a tropical forest rose on the land, and dinosaurs played in the mud. Then, still slowly, the land grew cold. Ice blankets thousands of feet thick pushed south from Canada, leveling this middle land. When those glaciers melted, they left thick, rich soil where prairie grass began to grow. Can you see the grass? Well, if you were down there you couldn't see over it; most of it is higher than your head.

Let's go on, over the broad river the Native Americans call Mississippi (which means "big river" in the language of the Ojibwa tribes). Some Indians call the Mississippi the "Father of Waters." It is the largest river in North America. Mississippi is easy to spell. Just say out loud: *M, I, double S, I, double S, I, double P, I.*

Did you ever before see a river from up high? Notice that it looks like a tree, with a fat trunk and many branches. A tree grows from its roots toward its top, but a river flows from its tiniest branches down to its base. The Mississippi has two huge branches: one, the Missouri River, flows from the west, the other—the Ohio River—flows from the east. Someday a big city, called St. Louis, will be built near where the branches come together. The M-I-double-S-I-double-S-I-double-P-I splits

The Missouri River is named after the Siouan (SOO-un) tribe that lived along its banks. But *Missouri* is not a Siouan word. How did that happen? The Illinois, an Algonquian tribe, named the river, using a word in their language that means "owners of big canoes."

the land in two—but not in half. The western part is twice as big as the eastern part.

Up there in the northland, near where the Father of Waters begins (it is called its *source*), it looks as if someone spilled five buckets of shining blue paint. Those five blue puddles are lakes—enormous lakes. If you were a fish swimming in the middle of one of them, you would think you were in the ocean. That is how large they seem. They are called Great Lakes, and that is what they are. (See if you can find their names and write them on a map. All are Indian names, except the largest. What does its name mean?)

We haven't finished this trip. We're still heading east, over rolling land and rivers, over the ancient Appalachian Mountains, over trees that are a thousand years old and seem to hold the sky in their branches. From the Mississippi east to the ocean, much of the land is covered with a carpet of trees.

And those trees, right now, with the first chill of fall, are like an artist's palette filled with brilliant reds and yellows and golds. Traveling across the land we have seen natural wonders: towering redwood trees, snow-topped mountain peaks, awesome stone spires in the desert, crashing waterfalls—but nothing is more remarkable than this wondrous display of seasonal color. Here the sky is so full of birds—thousands and thousands of them in flocks that stretch for miles, all heading south to their winter homes—that you need to watch the capsule or you could have a collision.

On we go. Now that we have cleared the Appalachians, there are smaller hills, called foothills, then rich, flat land dotted with Indian farms, and then low coastal lands with swamps and marshes where water rises and falls with the ocean tides until, finally, we reach the ocean itself, and the beaches it licks. Here, on the mid-Atlantic coast, a Chesapeake Indian girl and boy pick delicate sassafras plants. Their parents will brew the roots into a healing, pungent tea. Bright red berries dot the glossy green leaves of holly bushes here, and perfumed blossoms fill the branches of the shiny-leafed magnolia trees. (The boy picks one of those big, creamy blossoms and puts it on his sister's head; it fits like a cap.)

Here, by the sea, the aromatic leaf of the bay tree is prized for the flavor it brings to meat stews. Here wild roses flower and long-necked white cranes pick crabs from the shore waters. Someday, when sailors come from Europe, they will tell of smelling the fragrant land before they see it. What a place this is, and how fortunate are the people who live here!

In Europe, early illustrators of the New World had a hard time. Artists who couldn't go to America to see things for themselves had to guess how they might look from descriptions of explorers like Columbus. Sometimes they got it wrong. This is a sassafras tree—check an encyclopedia for a picture of the real thing.

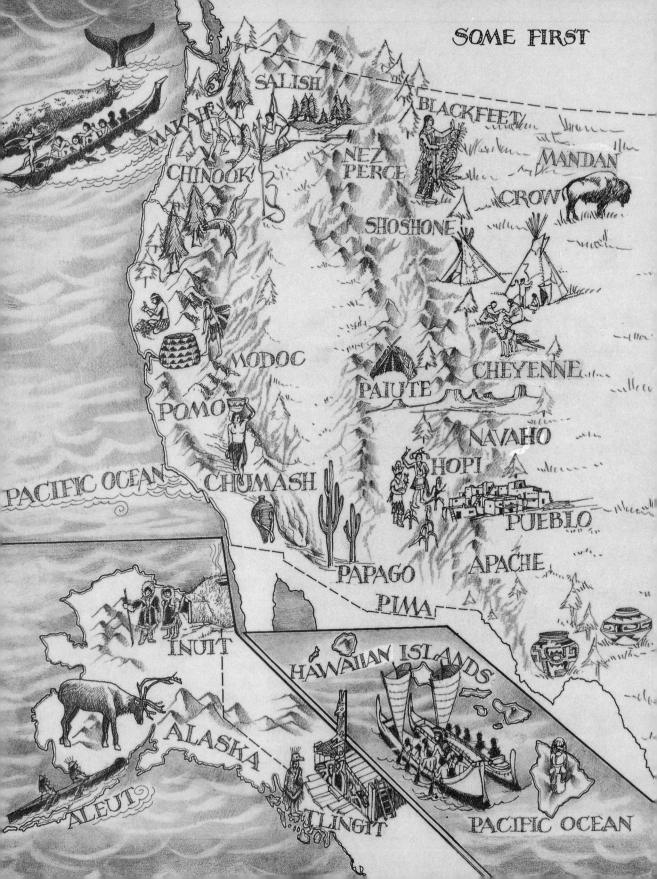

SOME FIRST

SALISH
BLACKFEET
MAKAH
MANDAN
CHINOOK
NEZ PERCE
CROW
SHOSHONE
MODOC
PAIUTE
CHEYENNE
POMO
NAVAHO
PACIFIC OCEAN
CHUMASH
HOPI
PUEBLO
PAPAGO
APACHE
PIMA
INUIT
HAWAIIAN ISLANDS
ALASKA
ALEUT
TLINGIT
PACIFIC OCEAN

AMERICANS and WHERE THEY LIVED

OJIBWA

MENOMINEE

SIOUX

SAUK

ILLINOIS

MIAMI

PENOBSCOT

MASSACHUSET

WAMPANOAG

IROQUOIS

NARRAGANSET

ERIE

DELAWARE

POWHATAN

CHEROKEE

WICHITA

CHICKASAW

CADDO

CREEK

COMANCHE

CHOCTAW

ATLANTIC OCEAN

TIMUCUA

GULF of MEXICO

9 Plains Indians Are Not Plain At All

This shield cover shows a Sioux warrior in battle.

Remember how the First Americans found a hunter's paradise in North America? Then the climate changed. Many animals, like the mammoth, became extinct. Much of the land dried up. It was even drier than it is now. Life became hard for the people who were hunters. They had to adapt, to change their ways. They had to hunt new animals, gather new foods, and learn to plant crops. Some Indians must have longed for the good old days.

Let's zoom back to the center of the continent. It is the year 1000. If you look out the window, you can see people gathering nuts and grasses and berries. It is summer and the sun is shining on the flat plains that make a wide ribbon across America. That ribbon stretches from Canada south to Texas, and from the Rocky Mountains east to the woodlands near the Mississippi River. Mostly it is windy land, and dry, with extremes of heat and cold. If we get out and stand for a while, you'll see for yourself the vast, open landscape. Your eyes will reach for miles and miles. Wherever you look, the sky touches the earth, and neither mountains nor forests block your view.

The people here live in a region with blizzards, tornadoes, icy cold, blistering heat, droughts, floods, and sometimes nice, balmy days—like this one. Some of the Plains land is prairie. Tall grasses grow on the prairie, and animals feed and hide in the grass. In some places, rivers web through the Plains. Where there are rivers there are trees and, often, Indian farms. But much of this land has only scruffy growth. There are few trees. Farming is difficult. Most of the people we see are nomads who keep on the move following the trail of the animals. It isn't easy, hunting on foot. These Plains Indians use their brains and their courage and stampede animals into traps. They use bows and arrows. All that helps, but not enough. These are among the poorest and hungriest of the Native Americans.

They live in tepees made of animal skins. It is the job of the Indian woman to put up and take down the tepee. She can do it in about an hour. If you watch, you'll see that in summer some tribes move almost every day. They knock down their tepees and load them onto two poles held with a crosspiece that forms an A. The wide end of the A drags on the ground, the other end is pulled by a dog. These Indians have few possessions.

Mostly they hunt buffalo. Buffaloes are fine for hunting for three reasons: they are good to eat and rich in protein; there are millions of them, perhaps 100 million in North America; and they are stupid animals. That's a nasty word, but it just happens that buffalo are not smart at all. That helps the men who hunt them.

While you are controlling the capsule, how about zooming ahead to the 16th century? Do you see those men down there, right where Dorothy and Toto will live one day? (That's Kansas, of course!) They are Spaniards. One of them, who travels with an explorer named Coronado, keeps a journal. In the journal he describes what the Indians of the Plains do with a buffalo after they finish eating the meat.

> With the skins they build their houses; with the skins they clothe and shoe themselves; from the skins they make ropes and also obtain wool. With the sinews they make threads, with which they sew their clothes and also their tents. From the bones they shape awls. The dung they use for firewood, since there is no other fuel in that land. The bladders they use as jugs and drinking containers.

Remember the buffalo corral in Chapter 8? Plains Indians were experts at trapping animals.

Sinews are tendons—tough, elastic strings—that connect muscle to bone. Animals have sinews, and so do we.

Awls are small tools for making holes in leather.

Dung is animal droppings.

Some people actually think that all Indians used to live in tents (called tepees). You know that isn't true. Some *did* live in tepees, but others lived in ice houses or wood houses or clay houses—or houses of thatched grass, like these in a Wichita village in the plains.

A warrior of the Mandan people. They hunted and farmed. They lived about where North Dakota is today, in round, earth-covered houses.

The best Indian

knives are made of a dark volcanic glass called *obsidian*. They are sharper than steel knives and keep their edges longer. But in the 16th century, Indians want white men's knives. Steel knives glisten; they are new, they are different, and they seem better.

Up in your capsule, you can see the pale-skinned men trading with the darker-skinned men. They give the Indians horses and knives. Soon those Indians will not be poor anymore. Horses will change their lives. (Think about facing a herd of buffalo on foot!) They will be able to gallop across the plains. They will hunt more efficiently. (They will also make war more efficiently.) They will travel great distances and trade with faraway peoples. They will feel free—and powerful.

Now, how about setting the controls for the 18th century? See those handsome Indians on horseback with feathered headdresses? They are Plains Indians. Some are people of the old tribes, and some are newcomers who have moved from the eastern woods. They have created a great, new Indian culture full of ceremonies, dances, horsemanship, costumes, warfare, and elegant crafts.

With horses and rifles (they trade buffalo skins for guns) these Plains Indians hunt and kill large numbers of animals. No longer do they have to use every part of the animal; now that they are affluent, they have become wasteful.

The Plains Indians are deeply religious; they have many traditions and strict rules of behavior. "Something happens to a man when he gets on a horse, in a country where he can ride forever," a historian has written. "…he either feels like a god or closer to God."

Now that they can hunt so efficiently, they farm less. Do you see those huge herds down there? Of course you do! You can't miss them. Buffalo—millions and millions of them—stretch for miles across the level land. The Plains Indians depend on the buffalo for food and clothing. It is hard to believe, but by the end of the next century, the 19th, the buffalo will be almost extinct, killed by wasteful people. (Who do you think the wasteful people might be?) And the Plains Indians will be in trouble, too.

Too bad we can't stay longer with these vigorous people. But, before we go, perhaps you can trade your pocket knife and get a feathered headdress or an otterskin vest decorated with porcupine quills or a rope of fragrant sweetgrass to freshen the air in this capsule. Then put your mind on the navigation controls, because we're going to zip around in time and space.

We're heading east, to the woods that stretch from the Mississippi River to the Atlantic Ocean. Now set the time dial to the beginning of the 11th century, which is where we were when we first met the poor people of the Plains.

Guess what the Indians who live in the eastern woodlands are called. Yes, Woodland Indians. You're going to be surprised when you see what some of the Woodland Indians are doing. Hint: think of the way ants carry dirt and build giant mounds. Weird? You'll see.

10 Mound for Mound, Those Are Heavy Hills

This gorget (GOR-jit) was carved from a shell, maybe for a priest. Can you see the raccoon tails?

Do you see the Mississippi down there? Now look up, down, around the river, and way to the east—to the mountains and even beyond. We're going to check out Mound Building Indians, and this is the region where they live. You will be awed: these First Americans have fine cities, well-organized governments, beautiful art objects, and successful businesses. It is their mounds, however, that you may never forget.

Mounds? Yes. Imagine thousands and thousands of Indians carrying baskets of dirt and dumping them to make hills. Skillfully shaped dirt mounds were all over the place when the white people first arrived in America—hundreds of thousands of dirt mounds. One of our presidents, Thomas Jefferson, dug into a mound. He dug carefully, because he had a scientific mind and he wanted to learn about Indians. But most mounds were just pushed down by farmers or builders.

What were the mounds for? Some mounds were graves. These Indians made a big ceremony of death, just as the ancient Egyptians did. (The pyramids in Egypt are burial mounds.) But some Indian mounds were used as platforms for temples and leaders' palaces, and some may have been religious symbols.

If we fly into the present time—right now—near Cincinnati, Ohio, we can see a huge curving snake. It is an earth mound that coils for a quarter of a mile. Amazing, isn't it? It may be more than 2,000 years old. What is the purpose of the snake? Or of a mound shaped like a turtle? Or of some of the other animal mounds? No one today is quite sure. Perhaps someone in your generation will figure it out.

Great Snakes

The Serpent Mound in Ohio was one of the United States' first ever conservation projects. It was discovered in the middle of the last century, just before the Civil War (do you know when that was?). A tornado mowed down the forest that had kept the mound hidden. Farmers wanted to plow it up. Some society ladies in Boston wrote to people about the danger the snake was in, and the people sent money to pay to save the mound as a national monument.

The great snake of the Serpent Mound is over 1,300 feet long. Its mouth is open and it is swallowing something—an egg, perhaps.

The Mound Builders' snakes and bears are not the only things of their kind. Other peoples have drawn or dug enormous pictures or monuments on the earth. Such designs can be seen fully only from up in an airplane. Why do you think people would make designs that they could never see properly themselves?

We know quite a bit about some Mound Builders because of a 19th-century farmer named M. C. Hopewell. Hopewell found 30 mounds on his farm in Ohio. He had archaeologists (ar-kee-AHL-oh-jists) dig carefully into them. Archaeologists are scientists who are trained diggers. From pieces of pots and bones and throwaway things they can tell a lot about the past. Historians would be lost without archaeologists.

The archaeologists on the Hopewell farm found more than just old pots and bones. They found copper, pearls, shells, mica, soapstone, and obsidian. They found teeth from sharks and teeth from grizzly bears. Most of these things had come from far away: the shells from the Atlantic coast, the obsidian from the Far West, the copper from mines near Lake Superior. So we know the Mound Builders were great

traders; we think they used a kind of relay system to get goods to and from distant places. Their sophisticated culture was flourishing at the time when Christ lived. (And when was that?)

Here is how an archaeologist who dug at Hopewell described a grave he uncovered:

> *At the head, neck, hips and knees of the female and completely encircling the skeleton were thousands of pearl beads and buttons of wood and stone covered with copper; extending the full length of the grave along one side was a row of copper ear ornaments; at the wrists of the female were copper bracelets. . .[she wore a necklace] of grizzly bear canines and [a] copper breastplate on the chest.*

Sophisticated (so-FIS-tih-kay-tid) means complex or worldly-wise.

Canines (KAY-nines) are teeth that are good for tearing food, especially meat. (Dogs are *canines*, and dogs eat a lot of meat.) Can you find out which of your teeth are the canines?

45

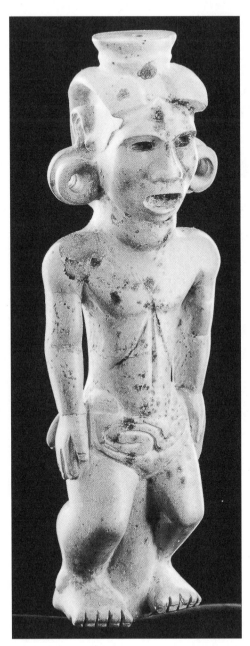

This little stone man is actually a pipe, over 2,000 years old. He and objects like him help us know how the Adena mound people dressed and adorned themselves.

Who was she, that Indian female? She must have been special to have that robe with its thousands of pearls and its bright copper buttons. She was buried with a man who was also covered with finery. Were they a royal couple? That is one of history's mysteries.

Now let's fly in our capsule, through a thousand years of time, to the year 1000, until we reach the Indian city of Cahokia (kah-HO-kee-ah). We are near three great rivers: the Mississippi, the Missouri, and the Illinois. This is a marvelous spot for a trading people to place a city. (Someday a city named St. Louis will sit across the Mississippi from here.)

Do you see the mounds? That's a foolish question—you can't miss them. One Cahokian mound is as tall as a 10-story building. Its base is broader than that of any of the pyramids in Egypt. The mounds look like flat-topped pyramids with temples and public buildings and statues on their summits. All those people in the streets are going to markets and schools and businesses.

Cahokia covers six square miles. About 25,000 people live here; another 25,000 people live in nearby villages.

Cahokia isn't a democracy; it is a slave society with a powerful ruler who is called the Great Sun. He is thought to be the earthly brother of the heavenly sun.

Slaves are called *stinkards*. They are captured in battles with rival tribes. Stinkards do work no one else wants to do (like hauling dirt to build mounds, perhaps). Noblewomen have to marry stinkards. Why? We don't know, but maybe it is a way to bring outsiders into the tribe. (If people keep to themselves, and don't ever marry outsiders, their children may be born less healthy. The science of genetics explains why.) Maybe it is a way to keep any one group from holding too much power. Maybe the stinkards are attractive. People from other places often seem more interesting than those you know well. The children of noble-stinkard marriages don't get stuck in the dirt: they go into an upper class.

Cahokia was big and thriving, and then something happened. It disappeared as a great city. Why?

Some experts think it may have grown too large. They say the inhabitants may have destroyed the nearby forests to get firewood, and without wood, their city couldn't survive. Maybe the sanitary system wasn't good and people got sick. Perhaps enemies attacked. Perhaps the people got tired of the slave society. No one knows for sure.

Digging Up Dirt on the Mound Builders

Historians like to divide the past to understand it better. You know: Stone Age, Iron Age, and so on. It helps to take long periods of time and make them orderly. That is why the Mound Building cultures—which flourished for more than 2,000 years—get divided into three periods. The oldest Mound Builders—who lived about 2,500 years ago—are called the Adenas (uh-DEE-naz). (They are named "Adena" for the region in Ohio where some of their mounds were found.) They built cone-shaped mounds for the dead and animal-shaped mounds that may have had religious importance. They were hunters and gatherers who lived on a land rich in fish, game, nuts, and berries. They grew some crops.

New people arrived on their land. Were there wars? Or was there peaceful acceptance of the new arrivals? We don't know, but life seems to have become more interesting. The Hopewell era—about 2,000 years ago—was a splendid time for Mound Builders. The Hopewell Indians (of course, they didn't call themselves that) lived in villages along thickly populated river regions. They traded with each other and with people across the continent. They were creative folk who left beautiful artifacts in their mounds. (Artifacts are people-made objects: stone carvings, jewelry, ceremonial pipes, hammered copper earrings, tools, and bowls.) Like the Adenas, the Hopewell Indians grew some crops. Their skeletons tell us they were very healthy and had few cavities in their teeth.

About 1,000 years ago the culture changed. Corn was brought from Mexico and the Mound Builders became serious farmers. With planned agriculture, cities could grow large. That brought highly organized governments. The rulers acted like gods. The mounds became complex engineering feats that took planned group effort. Temples and leaders' homes sat on top of the mounds. These Mound Builders are called Mississippians. Cahokia was their great center. The Mississippian diet centered on corn. Creativity declined. The Mississippians were less healthy than the Hopewell Indians had been. They were less free. Mississippian skeletons show tuberculosis and other diseases. Their teeth have cavities.

Was it because city life was stressful? Was it the limited diet? Or the dictatorial government? In the 1500s, when newcomers came from lands beyond the seas, the Mound Building culture was well past its peak. Diseases finished it off.

Iowa's Marching Bears form the second-largest group of effigy mounds in the U.S.

If You Would Like to See Mounds, Here Are Places to Go:

In Ohio—Serpent Mound is four miles northwest of Locust Grove on Route 73.

Mound City Group National Monument is three miles north of Chillicothe on Route 104.

In Indiana—Angel Mounds State Historic Site is in Evansville.

In Wisconsin—visit Aztalan State Park, between Milwaukee and Madison on Route 94.

In Iowa—Effigy Mounds National Monument is north of Dubuque on Route 76.

In Illinois—Cahokia Mounds is one mile from Collinsville, off Route 55.

Dickinson Mounds is between Lewistown and Havanna, off Route 78/97.

11 Indians of the Eastern Forests

Passenger pigeons weren't scared of people, so they were easily caught. Now none are left.

We're still zooming around in that time when the Native Americans shared two continents with no one but animals. We're flying low—between the Atlantic Ocean and the Mississippi River—and looking down at a sea of treetops. So thick are the trees that a squirrel might go from the Atlantic Ocean to the Ohio River and even beyond—jumping from tree to tree—and never touch the ground.

It is a fine place to be a squirrel, and not a bad place for humans either. The woods are filled with good food for animals and people. Passenger pigeons fly in formations so dense they darken the sky. When a flock lands, its weight bends trees low and men reach up and grab birds for their dinner. They are a treat, these passenger pigeons, sweet and juicy, and so abundant that the men net them and waste them. How can they know that someday the pigeons will be extinct and that most of the great trees will also be gone?

Some of these trees measure 30 feet around. Look at your belt. Now imagine a rope 30 feet long. Make it into a belt, and you get an idea of how big the trees are.

If we land the capsule and take a walk in the woods, you'll see that the Indians have cleared away the brush with fire, so the forest—with only tall trees and high grass—seems like a cool, sun-speckled park.

The openness of the forest, and the tenderness of the grass, make it inviting to animals. The woods are filled with beaver, deer, raccoon, possum, and bear. The hunters have an easy time of it. See? Over there? A hunter is wearing deer's antlers on his head. He walks softly in deerskin moccasins, pretending to be a deer himself. He is teaching a boy some tricks with the bow and arrow. The boy is his son, and they will have a good day and drag a heavy deer home with them.

A warrior of the Huron tribe in fighting gear. The Hurons, a group of four nations in the northeast, speak the Wyandot language. The Hurons and the Iroquois were enemies.

These Indians practiced a kind of farming called *slash-and-burn.* First they would clear most of the trees from a piece of land. Then they burned the branches and leaves. The rich ash was hoed into the ground to fertilize it. Then the land was planted.

They are Woodland Indians, as are all the Indians of this region; some are grandchildren of the great Mound Builders. They have heard stories of a glorious past, of temples and pyramids. But now the tribes are small and so are their mounds. Still, life is good. The men hunt in the woods and fish in the streams. The women who wait for them at home are farmers who grow corn and beans, squash and pumpkins. The women and children gather wild grapes, pick nuts from the trees, and sometimes dig clams on the beaches.

Our hunter's wife, who is mother to the boy, is the best cook in their village. She knows forty different ways of cooking corn. She will stew the deer meat and season it with the vegetables she grows.

The men have cleared trees so the village can sit in the sunshine with open fields for growing crops. They get rid of trees by "girdling" them. That means they cut the bark all the way around the trunk. That kills the tree, although it takes some time for it to die. When it happens, the tree falls down and can easily be split for firewood.

Now that our hunter is home, he has taken the antlers from his head. He isn't wearing much else, just a strip of leather that goes between his legs and hangs, front and back, from a belt. In winter the

The Woodland

Indians knew how to combine foods in nutritious ways. Beans, corn, and squash give more protein when they're eaten together than by themselves. So these forest Indians invented the dish of mixed vegetables that is still called by its Algonquian name—*succotash.*

49

Some Europeans who came to America in the 1500s and 1600s painted Indian ways of life. Artists like Englishman John White and Frenchman Jacques LeMoyne are important because their pictures show how Indians lived before foreigners changed things forever. LeMoyne made this picture of deer hunters in the early 17th century. Do you think the deer were fooled by the men under the deerskins?

hunter will add a shirt and leggings—both fashioned from animal skins. On special occasions he will bedeck himself in a fancy robe of turkey feathers. His wife wears a wraparound skirt made of deerskin. His son, who is the oldest child, dresses as his father does. What do the small children wear? Why, nothing at all. (In winter, too? Of course not!)

These people, who live in the warm South, are like people everywhere: they care about the way they look. Their clothing may be simple, but their makeup and jewelry are elaborate. Tattooed designs cover most of their bodies. Today our hunter will paint himself with bright colors to celebrate the hunt. He makes his skin glisten by rubbing it with bear fat. Because he is handsome and a bit vain, he will blacken his teeth with tobacco ash. It is the fashion.

He grooms his hair carefully. People across the world are inventive with hair, and the First Americans are no different. Our hunter shaves his head, with a sharp shell, leaving an island of hair on top to which he ties feathers. He leaves another island over one ear and makes a

Like Indian languages, some European languages are related to each other. French, Italian, Spanish, and Portuguese have a Latin base and many similar words.

thin braid with those hairs.

The hunter wears strings of pearls around his neck. His bracelet is of polished deer ribs that were bent and shaped in boiling water.

He and his family live in a one-room house made of narrow tree limbs lashed together with vines and covered with bark. It is called a wigwam. The house has a round roof thatched with strong reeds. It is similar to the other hundred homes all clustered around a central square. Do you see those children turning cartwheels down the village street? And the others playing a game of ball? It would be fun to stay and watch the game, but we need to head north.

From the capsule we can see Indian villages dotting the forests and coastal lands. Most of them belong to Indians who—like our hunter and his family—speak Algonquian languages. That doesn't mean all these villagers speak the same language. It means their languages are related—and most of their customs, too. Many of the Algonquian tribes trade with each other and are friends.

They have enemies who speak a different language. The Algonquians call them "terrible people," or "frightening people," or sometimes "rattlesnake people." In the Algonquian languages the name of those enemies is Iroquois (EAR-oh-kwoy).

They are unusual, these "terrible" people. They believe in peace and brotherhood, but when they fight they are fierce and cruel. Their name for themselves is Haudenosaunee (ho-dih-no-SHAW-nee), which means "people of the long house." It is a descriptive name. Their houses are long, sometimes 150 feet long—sometimes longer. The longest—said to be 334 feet long, even longer than a football field—was built by men of the Onondaga tribe.

Do you see that longhouse below? Twenty families live there. Actually, they are all one big family headed by a grandmother, with brothers, sisters, aunts, uncles, cousins, and other relatives. (No one ever has to hire a babysitter.)

The Iroquois have formed a league of Indian nations and wish to bring the Algonquians and other Indians into that league. They say it is a league of peace. The Algonquians want no part of the league. They do not want an Iroquois peace. They don't want anything to do with their ancient enemies.

The Abenaki, Ojibwa, Mahican, Massachuset, Narraganset, Powhatan, Blackfoot, Delaware, and Cree are a few Algonquian-speaking tribes.

We're going to use the name *Iroquois* for the Haudenosaunee—it is the name most used today. Remember, names can get chosen for the wrong reasons—then they stick. *Iroquois* has become a name the Haudenosaunee use proudly.

The Iroquois were not the only Indians to form a league. The Hurons did, and in the south, the "C" nations—Creek, Choctaw, Cherokee, Chickasaw—had an ancient bond.

12 People of the Long House

The artist has drawn these longhouses a little small. They could measure over 150 feet.

The Iroquois have no written language; that doesn't mean they have no history or government or records. Chiefs called *sachems* (SAY-chums), and other leaders of long memory and wisdom, are keepers of the people's past. They have a memory aid—a kind of picture writing—done with thousands of tiny shell beads. The beads, called *wampum,* are strung on leather cords and woven into designs. Often they are sewed on deerskin belts. Wampum is very valuable to the tribes and is sometimes used as money. The designs tell stories. Some record treaties; some tell the history of a clan.

The Iroquois have much history to remember and a remarkable form of government. Here in America, in the 16th century, the Iroquois have fashioned a democratic league of five Indian nations (a sixth will be added) with leaders who are expected to serve the people. It is a confederacy (kon-FED-er-uss-ee), which means each of the nations has its own identity and laws, except in matters of war or common concern. In those cases, a council of all the tribes makes decisions. All this has been organized in a plan of government that has been woven in wampum, memorized, and told by the sachems.

Fifty male sachems, ten from each nation, sit on the council. The sachems are chosen by women who head family clans. In the Iroquois world, women are much respected. Perhaps that is because the Iroquois depend on their farm crops, and it is the women who do most of the farming. The Algonquian peoples are hunter-gatherers who do some farming. These Iroquois are farmers who do some hunting and gathering. The Iroquois women are excellent farmers.

The Iroquois have a matrilineal (mat-truh-LIN-ee-ul) society. That word begins with a Latin root. In Latin the word for "mother" is *mater.* In a matrilineal society your descent is traced through your mother. Iroquois women are leaders of family clans. Our society in America today is patrilineal: our names usually (but not always) come from our fathers (what do you think *pater* means in Latin?).

The Iroquois men are excellent talkers. There is much discussion when the council meets. The Iroquois speak eloquently and listen carefully. The sachems persuade with words. All council decisions must be made unanimously—that means everyone has to agree. Sometimes that takes much talking.

In the days of dim memory, before the league was founded, the five nations fought each other in terrible rounds of anger and revenge. Some warriors were cannibals; they threw their dead enemies into pots and ate them.

Then three men appeared. Two were good but seemed powerless;

Henry Wadsworth Longfellow wrote a famous poem called *Hiawatha.* Unfortunately, Longfellow got his history mixed up. His story is about a Chippewa Indian. It has nothing to do with the real Hiawatha.

one was an evil ruler. Their names were Deganwidah (d a y - g a n - W E E - d a h), Hiawatha (hi-uh-WAH-thah), and Tadodaho (tuh-doe-DAH-ho). They were real men, although stories have been told about them that are myth.

Deganwidah was a poor, fatherless boy who could hardly talk. He stammered and stuttered each time he tried. This was a terrible problem, because Deganwidah believed he had been born to bring peace to all the Indians. So when his own people—the Hurons—wouldn't listen to him, he went to the Iroquois and preached, as best he could, of peace and brotherhood.

Now it happened that, among the Iroquois, there was a warrior named

An Indian man and woman are eating their food, a dish of boiled corn kernels. Still to come are fish and more corn; they like clams and nuts, too. The artist who drew this wasn't very good at showing the way Indians looked. His Indians seem more like a European idea of handsome men and beautiful women.

Hiawatha who was tired of killing and cannibalism. He, too, spoke of peace. He, too, wanted to end the bloody wars of revenge. But a terrible tyrant, Tadodaho, opposed him. When people began to listen to Hiawatha, Tadodaho was enraged. One by one, Tadodaho had Hiawatha's three daughters killed.

Hiawatha was in despair, and then he met shy, greathearted Deganwidah. Deganwidah had a plan for uniting the tribes into a peaceful league. He had a plan for ending the blood revenge and bringing evil-doers to justice. But he couldn't talk well. Hiawatha, however, was an awesome speaker. They went from tribe to tribe with a message of peace and unity.

But what should they do about the terrible Tadodaho? (The myths say Tadodaho had serpents growing out of his head.) Should they try to fight him? The tribal leaders all feared Tadodaho; they knew there could be no peace unless he joined them. Deganwidah sent Hiawatha —whose name means "the comber"—to speak to the tyrant. We don't know what really happened, but Hiawatha seems to have converted Tadodaho to ways of peace. (Myths say he combed the snakes from Tadodaho's hair.) Perhaps it was Hiawatha's gift of oratory (ORR-uh-tor-ee), or maybe he did some political bargaining. We do know that when the former tyrant joined the League of the Great Peace, he was given its most important role. He was named "firekeeper" of the confederacy, which means he was chairman of the council. Today the man who holds that position is still called Tadodaho.

When did all this happen? No one is exactly sure. It was before— but not too long before—the first Europeans came to the land of the People of the Long House. By the time the new Americans arrived from across the great ocean, the Iroquois were a powerful nation.

When the Europeans came, they called North America a "New World." It was a new world to them.

To the Indians it was not a new world, it was their old world; it was their home.

At first some Indians tried to be friends with the newcomers, and some of the newly arrived men and women returned the friendship. But many people on both sides understood that these two civilizations were very different. They would have a difficult time living together on the same land. And they both wanted that land.

Most of the Europeans called the Indians "savages." They thought they were wild, uncivilized creatures, almost animals, without proper customs or organizations. They didn't understand that the Native Americans had religions, governments, and ideas of their own. The Europeans read a Bible that told of Abraham and Moses and Jesus. The Iroquois had their own creation stories. They called North America "Turtle Island." Their stories told them that Sky Woman fell through a hole in the sky and Great Turtle caught her. Land began to grow at her feet. That is how the world began, they said. The Bible said the world began with Adam and Eve in the Garden of Eden.

The Iroquois thought themselves the most powerful people on

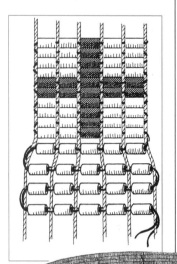

This wampum belt may tell the story of Deganwidah, Hiawatha, and Tadodaho. The picture above it shows one way of weaving wampum.

The constitution of the Iroquois Confederacy was said to have been woven into wampum about 1570. Its three founders may have lived a hundred or more years earlier.

This village of longhouses and gardens looks very settled and orderly. But after a dozen years or so, the soil in the fields will lose its goodness. Pests such as caterpillars and centipedes will become a problem. The supply of trees for buildings and firewood will run low. So the villagers have to move—though probably not far away.

One Nation

When European men and women arrive in the Northeast, they learn of the Iroquois League. Men like Benjamin Franklin and Thomas Jefferson will understand that the Iroquois tribes form one nation, but each also keeps its own identity. The American states will do the same kind of thing.

Turtle Island—and Turtle Island was the only world they knew. The Europeans also believed themselves the most powerful people in the world, until they learned of China and Japan. Then they weren't quite sure. Some of them were on their way to find out about those people of Asia when they ran into Turtle Island. As you'll soon see, it was a surprise.

13 Let's Turn North

In a small boat, it was easy to believe that a big storm was a sea monster.

The first Europeans to discover America came from the lands of the north. Called Vikings or Norsemen, they were the terror of Europe. Their ships were fast, their seamen brave and blood-thirsty. Thor, Odin, and Loki were some of the gods they worshiped. They sacrificed animals to those gods, and sometimes they even sacrificed humans. Vikings told stories of elves and giants and trolls and believed they really existed. Norse stories are among the best you will ever read.

Viking means "sea raider" or "pirate," but not all Vikings were pirates. Most were farmers who kept cattle and sheep. Historians think that during the 9th and 10th centuries their homelands in Scandinavia became crowded, so Vikings set out for other places. Some sailed in search of loot, but others looked for fair lands and good fishing.

(This is a good time to check a map. Scandinavia is northern Europe: Norway, Sweden, Finland, and Denmark.)

The first Viking ship arrived in America by mistake. Bjarni Herjolfsson (bee-YAR-nee HUR-yolf-sun) was a Norse sailor on his way to Greenland when the wind blew him off course. We think that was in the year 986. He went home and told his people what he had seen. One of his friends, Leif Eriksson, decided to explore the new land. Leif (LEEF) was called "Leif the Lucky" and was the son of Erik the Red, a famous explorer who had discovered Greenland (and had red hair).

Erik was fierce. He had gone exploring because he was wanted by the law. But Leif, according to some old writings, was a "fair-dealing man." Leif was "big and strong" and a great sailor. He followed Bjarni's route and landed in a place he named Vinland because it had wild vines. Vinland was probably Nova Scotia (SKOE-shuh), which is now

Loot is stolen goods, also called *booty*.

Greenland, far to the north between Canada and Europe, is the world's largest island. It is still in the Ice Age. Much of the island is covered by a glacier that is two miles thick.

part of Canada. (The vines were probably berry bushes, not grape-vines as people once thought.)

Leif's brother, Thorvald (TOR-vult), was the next Viking to come to America. He intended to stay. One day, while he was walking over the land looking for a place to settle, he spotted three mounds on the beach. They turned out to be three canoes. Under each canoe three men were sleeping. They were Native Americans. Thorvald killed them all, except one who escaped. The next day a large party of Indian men came in canoes and attacked with bows and arrows. Thorvald became the first white man to be killed by Indians. What if he had tried to make friends with the Indians? Do you think they would have made friends with him? Would history have been different?

Thorfin (TOR-fin) Karlsefni was a Viking who did settle in America —at least for a while. He was married to beautiful Gudrid (GOOD-rid),

A small sailing ship can travel perhaps 70 miles a day in good weather with a strong wind. Can you imagine crossing 3,000 miles of ocean in a little Viking ship with one sail?

57

Norsemen were afraid of terrible sea monsters, such as this man-eating serpent. Most seagoing peoples believed in such things. The Norsemen believed in other worms, too. Norse legends say the earth is supported by a great tree. A huge worm or serpent is always gnawing at the tree's roots, trying to kill it.

Business Deals

At the beginning of spring they saw one morning early a fleet of skin canoes...Karlsefni and his men raised their shields, and they began to trade: the people wanted particularly to buy red cloth, in exchange for which they offered skins and gray furs. They wished also to buy swords and spears, but Karlsefni and Snorri forbade this.

—FROM *ERIK THE RED'S SAGA*

and she was the widow (the wife of a man who has died) of another of Leif's brothers. Thorfin and Gudrid came in three ships filled with cattle, sheep, and other Vikings. It must have been an adventure. Perhaps they felt as space explorers do today. They stayed for three years, and when they left they took their son, Snorri, back home with them. He was the first white baby born in America.

How do we know all that? Historians have many ways of finding out about the past. Like detectives, they search for clues. They use tools to help them solve mysteries.

Archaeology is one tool.

Literature is another.

Anthropology is another.

Zoology is one more.

Archaeology is the digging-up science. (Remember the mounds of the Hopewell Indians and the archaeologists?)

In 1961 archaeological diggers found a Norse settlement on the northern coast of Newfoundland. It proved that Vikings once lived in America. The archaeologists dug up part of a spindle for spinning wool. That was a clue. Now they knew that women had lived in the settlement. (Why would a spindle tell them that?)

The archaeologists found the remains of eight Viking-style long houses. From those remains they could figure out exactly when people had lived in the settlement. Radiocarbon dating told them. This is how it works. All living matter contains radioactive carbon 14. When a plant or animal dies, the carbon 14 starts to disappear. It disappears at a steady rate, and we know exactly how fast that happens. The amount of carbon left in an object tells us its age. Radiocarbon dating showed that the Viking houses were built soon after the year 1000.

Sometimes archaeologists dig up old bones and bits of pottery. Those bones and the way they are buried can tell us much about the way people lived. We have found mammoth bones together with Indian bones and spearheads in North America, so we know Indians hunted mammoths. So far we have not found Viking bones. Can you guess what old pieces of pottery might tell?

Literature helps with clues about history. Ancient documents found in Wales (which is part of Great Britain) tell that Prince Madoc of Wales, in 1170, sent 10 ships sailing west. Some say the sailors settled in a land they found across the waters. Do you think it was America?

A very old book, *The Saga of the Greenlanders*, relates the story of Bjarni and Leif and their voyages. It gives exact dates. The dates match the archaeologists' findings.

Anthropology (an-thruh-PAHL-uh-jee) is the science of people and how they develop. Anthropologists study fossils and living people. They study history, languages, and the ways people live. Anthropology is very helpful in understanding the past.

Zoology (zoh-AHL-uh-jee) is the science of animal life. Can that help historians? You bet. When diggers found a strange sea snail in the Viking settlement, they gave it to zoologists to study. The zoologists said the snail was a native of Scandinavia: it had never been found in America before. They believe it must have come over as a hitchhiker in the damp bottom of a Viking ship. It was another clue to prove the Vikings had been to America.

Sometimes historical clues turn up in strange places. In 1957 a bookseller brought a book to Yale University, in New Haven, Connecticut. The book had been hand-copied in 1440 in Switzerland, in the days before the printing press was invented. A map in the book showed a place called Vinland and said it was "discovered by Bjarni and Leif." Now we know there was a map of the New World before Columbus sailed.

Fossils are old bones or traces of ancient plants or animals. Sometimes people with old-fashioned ideas are called fossils.

Near the middle of the Vinland map are Scandinavia, England, Ireland, and the rest of Europe. Up near the top, toward the left, is Greenland. The big island on the far left (it isn't really an island) is Vinland, or Nova Scotia. Scholars still argue about whether the Vinland map is a fake.

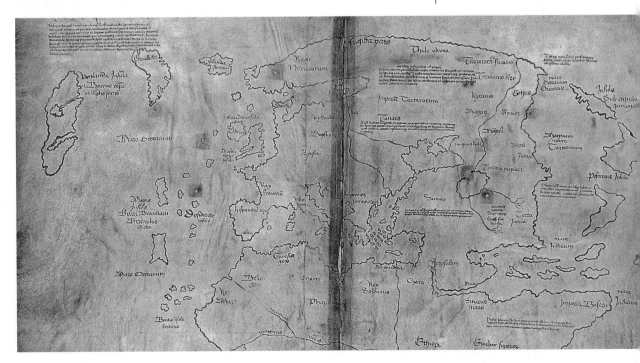

How Do You Know What You Know?

Now we know there was a map of the New World before Columbus sailed." That is the sentence that ended the last chapter. But do we really know that? Watch out for the word *know*. All through history people have been sure they "knew" the truth, but often they were wrong.

For a long time men and women "knew" the world was flat. After all, that was what their eyes told them. Everyone laughed at the first people who said the world was round. "What a silly idea!" they thought. Sometimes they persecuted the "round earth" thinkers for teaching dangerous ideas. But the "flat earth" people were wrong.

In the old, old days almost everybody believed in witches. Then people

learned there really weren't any witches, so they tried to kill everyone who might be a witch. Read that sentence three times, and see if it makes sense. Sometimes history is like that. Sometimes people just don't want to believe new "truths." Sometimes people are afraid to change their ideas. Sometimes they are just lazy; old ideas are comfortable and familiar. If you keep reading *A History of US,* you'll learn about witch trials in Salem, Massachusetts. You'll see they didn't make sense either.

That's one problem of human nature that history shows: when people think they are sure of something, they stop asking questions. That can be dangerous. Don't ever be afraid to ask questions.

Now back to the Vinland map. All the experts at Yale University agreed that the map was the real thing. The paper was old, and everything else about it checked out.

Well, not to everyone's satisfaction. One expert was bothered by the ink. He said the ink was not the kind used in the 15th century. He said the map was a fake. So now no one is quite sure. Scientists will have to keep studying the map in order to decide.

Many historians say that after the days of Thorfin and Leif the Lucky, the Norsemen stopped coming to North America. They say the

New World became a half-forgotten memory.

Other historians think they are wrong. Clues seem to say that the Norsemen kept coming to the new land in order to get furs and timber and fish. Vikings may have settled in New England and explored much of the country. No one is quite sure of that either.

Detectives would say that a stone found in Minnesota is a good piece of evidence. It has a message in early Norse writing. The first expert who looked at the stone—it's called the Kensington stone—said it wasn't real. He said it was the work of someone who was trying to fool the experts—and he certainly wasn't going to be fooled.

But now a few experts say it really is genuine. If they're right, that means the Vikings made it all the way to Minnesota—unless someone carried the stone there later. Look at a map. Could Vikings in boats have gone that far? Remember, don't be too sure of yourself!

Other clues have turned up in other places. When explorers came to Canada in the 16th century, they found Indians with blond hair and blue eyes. Most Norsemen had blond hair and blue eyes. Do you think the blue-eyed Indians could have been descendants of Vikings?

Historians, archaeologists, anthropologists, and other scientists still have a lot of work to do to solve all the mysteries of the Vikings in America.

This is the "Kensington stone" from Minnesota. The Norse letters carved on it are called runes. We still don't know if it's a fake.

14 The Power of the Press

An *armillary sphere* was like a globe of the sky. It showed the stars' position.

We need to get something straight—right now, before we go on with this book. It has to do with the centuries. You need to be sure you understand about them. Do you know that when you see "12th century" it means the years that begin with 11? The numbers are always 100 years behind the centuries.

It's the same when you have a birthday. Your birthday celebrates the number of years you've already finished. You may say, "I took a trip when I was nine," but you actually took the trip in your tenth year. (That's because after your ninth birthday you start on your tenth year in the world.) It's a bit confusing, but once you have it, it's easy. If something happened in 1454, it happened in the 15th century.

And something did happen in 1454—something big. A German goldsmith named Gutenberg (GOOT-en-burg), Johannes Gutenberg, printed a beautiful book, a Bible.

That may not seem like a big deal, but it was. Gutenberg had invented an efficient way to use a printing press. He did it with movable type—letters that could be used over and over again. (Actually, the Koreans and the Chinese had been using movable type for centuries, but we don't think Gutenberg knew that.) In Europe, before 1454, if you wanted to print a book you had to carve each page on a separate woodblock. That wasn't easy, especially as everything had to be carved backward. So most books were copied by hand. Think about that.

Then think about a big book like a Bible—or any big book. Guess how long it would take to copy it. Think about illustrating it. How much might each book cost?

Do you think you'd be doing much reading if you lived before Gutenberg and his printing press? Of course not. Chances are no one would teach you to read. It wouldn't be worth it. Unless you were very

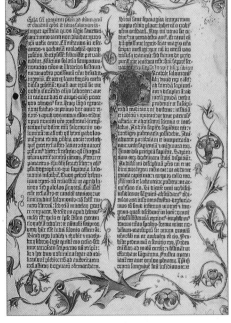

A page from the Gutenberg Bible. After Gutenberg, the way books were printed changed very little for 500 years. Then, not long before you were born, two machines—the camera and the computer—helped change printing again forever.

61

This is a page of a medieval (med-dee-EE-vul) manuscript, illustrated, or "illuminated," by hand. (The word *manuscript* comes from the Latin words for "hand" and "writing.") On the right is a detail from the same picture: it shows the scribe himself, doing his job.

How do books give power to people?

rich or very lucky, you might live a full life and never even see a book.

If you never saw a book or a newspaper or a magazine (and of course there was no TV), you wouldn't know much. You'd have to believe whatever you were told—by the lord of the manor or the king or the priests. It would be hard to think for yourself. A famous Englishman, Thomas Carlyle (he was a historian), said that Gutenberg's press created "a whole new democratic world."

And it was about time. For most people, before the 15th century, Europe was a place of superstition and poverty. For most it was a time of war and disease.

Then things began to change. In Italy, poets and painters and sculptors began creating new works of art. It was called a Renaissance (REN-uh-sahnce), a time of rebirth. Ideas seemed to be in the air, and inventions, too—like the compass.

Actually, the compass had been around for thousands of years. The ancient Chinese discovered that a magnet, swinging freely, will always point north. Arabs brought that knowledge to Europe. But early compasses were not always reliable. In the 15th century (the years beginning with 14) the compass was improved; it could now be depended upon at sea.

Having a little needle that always pointed north meant new worlds could be discovered. Imagine you're in a small ship in a great ocean. You can't see land—just water in every direction. If you don't have a compass, how do you know which way to go to get home?

If you know the stars, and most good sailors do, you can wait until nighttime and let the stars guide you. But suppose—just suppose—

it's cloudy and stormy. No stars can be seen. Maybe it's cloudy for a week. Your little ship can't carry much food. You might sail in the wrong direction and run out of food before you find your way back—if you get back. Storms at sea are tough to survive. All of which explains why ships stayed close to home before the compass was perfected.

Now we're still in the 15th century when, just as in the fairy tales, a prince appears: Prince Henry of Portugal, also known as Henry the Navigator.

A navigator is someone who knows where he is going: a kind of supersailor. Prince Henry never went far himself. But he was fascinated with sailing and mapmaking, and he inspired and encouraged others. Because of Henry, sailors, mapmakers, scientists, and mathematicians came to Lisbon, Portugal's capital. Everyone interested in exploration and the new sciences wanted to be there.

Prince Henry was a born explorer. He had an inquisitive mind. That means he was curious about the world around him and wanted to learn as much about it as he could. He also wanted power and riches for his country. He wanted his sailors to sail to China and Japan and India because they were thought to be the world's most advanced civilizations, and because they held gold and jewels and spices.

He wasn't the only one who wanted to get there. Most Europeans were wild to find a way to reach the Indies. (*Indies* was a catchall word for all the lands of East Asia.) And mostly because of a book. Almost everyone who could read had seen it. (At first there were only handwritten copies but then, thanks to Gutenberg, there were lots of copies.) The book was by Marco Polo, and it told of his trip to China (back in the 13th century). It told of golden palaces and jewels and wonders beyond imagining. Polo had done some exaggerating—still, people who knew him had seen jewels, remarkable jewels.

Whoever could find a fast, safe way to get to the Orient (another word for East Asia) would become rich and famous. Everyone agreed about that. But the only way to get there was by land, through Turkey and the Middle East. That route had become dangerous: rival Islamic empires—Ottomans, Turkomans, Mamluks, and Timurids—were fighting for control of the region. In addition, there were thieves who preyed on merchant caravans. (That doesn't mean they got on their knees to God. If you want to say that, you spell it *prayed*. If you spell it with an *e*, it means they robbed the caravans.)

Prince Henry was determined to have his sailors get to Cathay (China) by sailing around Africa, and finally the Portuguese did it. But one man dreamed of reaching China and the Indies by another, even faster, route. His name was Christopher Columbus.

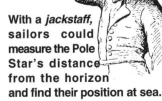

With a *jackstaff*, sailors could measure the Pole Star's distance from the horizon and find their position at sea.

15 A Boy Named Christopher Has a Dream

A fully rigged 15th-century ship called a *caravel*, much like Columbus's *Niña, Pinta,* and *Santa María.*

When Christopher Columbus was a boy he had two dreams. One was to go to sea; the other was to get to China. When he grew up, he thought he had done both.

Columbus was born in Genoa, an Italian city on the Mediterranean Sea. Genoa was prosperous because of the sea trade, and many Genoese boys wanted to be sailors. Columbus became one of the best the world has ever known.

Columbus knew about China because he had read Marco Polo's book. He had read it carefully. His copy of the book is full of notes.

Most people in the 15th century couldn't read. That meant they were ignorant of many things. Some believed the world was flat. They thought if you sailed too far you'd fall off the edge. But people who could read, like Columbus, knew that wasn't so. Scientists had proof that the world was round, and they told about it in books.

There was a problem, though. No one was quite sure how big the world was. So no one knew how far you would have to sail to go around it. One way to try and figure that out was by measuring lines of longitude and latitude.

Longitude and latitude are very useful lines. How about looking at a map—right now? That's the only way you will understand what is coming next. Do you see the thin lines going up and down and across the map? Those are lines of longitude and latitude. They are imaginary lines—you won't see them if you look down from an airplane. They are drawn on maps to help map readers divide up the globe. Latitude and longitude lines make it easy to read a map and measure the earth.

Columbus first went to sea at age 14. When he was 26, he was shipwrecked off Portugal and swam to shore. He sailed to Iceland, too, and may have seen Norse maps.

Columbus's brother was a mapmaker. That was a big help to Columbus.

Think of the earth as a big, fat man. Put a belt around his middle. That belt is a line of latitude. We call it the equator, or zero degree line of latitude (0°). Latitude lines are numbered north and south of the equator. Now give the fat man a round cap. The edge of the cap is the Arctic Circle, which is the 66½ degree line of latitude north (66½°N) of the equator. The center of the cap is the North Pole (90°N). Turn the fat man the other way, and the cap on the globe becomes the Antarctic Circle; now it has the South Pole as its center. (And now those numbers are 66½°S and 90°S.)

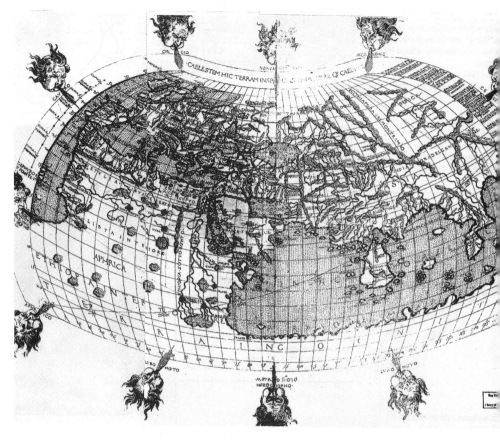

Lines of latitude circle the globe and run parallel to each other. Some people even call them *parallels*. (Parallel lines are an equal distance from each other and never touch, like the sides of a ladder.) Lines going the other way—from the North Pole to the South Pole—are lines of longitude. Longitude lines are not parallel. They all touch at the poles but spread far apart at the equator.

One more thing: those lines of longitude and latitude are actually circles. They circle the globe. If you divide the earth at the equator—that zero degree line of latitude—you get two halves. Those halves are called hemispheres. (Another word for a globe or ball is a sphere. Half a sphere is a hemisphere.) We live in the Northern Hemisphere. If you divide the world in half on a line of longitude, you will also get two hemispheres; this time they are Eastern and Western Hemispheres.

To tell longitude from latitude, remember that the first syllable of LATitude rhymes with FAT—like the belt around our fat earth.

This map formed Columbus's idea of the world and how it looked. Ptolemy, the Greek geographer and mathematician who first drew the map, lived in the second century. So his map was very old to Columbus—much, much older than Columbus is to us.

Compare this map to a modern map to see how mixed up Columbus was.

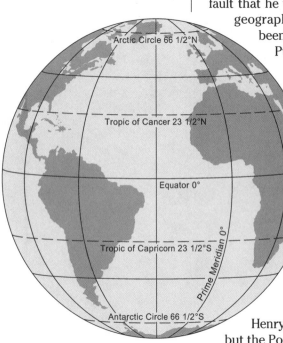

Arctic Circle 66 1/2°N

Tropic of Cancer 23 1/2°N

Equator 0°

Tropic of Capricorn 23 1/2°S

Prime Meridian 0°

Antarctic Circle 66 1/2°S

The lines that run horizontally (across) the globe show latitude. The vertical lines (up and down) show longitude. To tell where you are on earth, you need two numbers: a latitude and a longitude. Virginia Beach, Virginia (where I live), is at about 37°N (latitude) by 76°W (longitude).

LONGitude starts with "long"—like the long distance from the North Pole to the South Pole.

Understanding latitude and longitude can help you figure out a lot of things. Columbus, by the way, was very smart. He did everything well, except one thing. When he measured the earth he goofed.

He figured the earth was much smaller than it is. He also figured that Cathay was much larger than it is. Actually, it wasn't Columbus's fault that he was mixed up. He studied the work of an ancient Greek geographer named Ptolemy (TOE-luh-me), and if Ptolemy had been right, China would be where America is. If you compare Ptolemy's map with a modern map, you will see just how mixed up Columbus was.

Now as you know, people who could read knew the world was round. They understood that if you went west from Europe you would finally get to Asia. But no one wanted to try going that way—it seemed too dangerous and too far. Most people believed there were ferocious monsters in the deep waters. Every sailor knew the dangers of storms at sea.

Because Columbus believed the earth was small and because he was a superb sailor, he thought he could make it to Cathay. Since he knew about latitude, he knew that the same line of latitude passed through Spain and Japan. And Japan (he called it Cipango) was right next to China! It would be easy to follow that line, wouldn't it?

So he went to Portugal to ask for help. Remember Prince Henry and his mathematicians? Well, Henry had died in 1460, but the Portuguese were still world leaders in exploration and navigation. When the Portuguese mathematicians figured out the size of the earth, it came out much bigger than it did for Columbus. Today we know they had it just about right. But, back then, nobody was sure because nobody had actually sailed around the globe. Anyway, the Portuguese weren't willing to take the risk. They turned Columbus down.

So did almost everyone else. Columbus took his ideas to one person after another. Each one said "sorry," except King Ferdinand and Queen Isabella of Spain, who said "perhaps." Maybe they were just being polite, because they seemed to forget all about Columbus. Years passed. He asked them again; this time they said "no." One thing you can say for Columbus: he never gave up. He was on his way to see the King of France when a messenger called him back to Spain. Finally, Ferdinand and Isabella had agreed to help. They gave him three small ships and some sailors, and sent him in search of China and Japan.

16 A New Land Is "Discovered"

None of the many portraits of Columbus was made during his lifetime, so no one knows what he really looked like.

If you've ever seen a painting of Christopher Columbus, forget it. All of them were done long after he died. We can trust descriptions of him written by those who knew him best: his sons.

They tell of a man who is six feet tall, slim, with blond hair that turns white when he is 30. He has the manners and dignity of a nobleman, although his father was said to be a weaver of wool. Perhaps if Columbus were less of a gentleman, he could better handle the rough men who sail with him. That will always be a problem for him.

It is August 3, 1492, and three tiny ships—the *Niña*, the *Pinta*, and the *Santa María*—set sail from Palos, Spain. Columbus, on the *Santa María*, is 41 years old and commodore of the three-ship fleet and its crew of 90 men. In his pocket is a letter from King Ferdinand to the Grand Khan, the ruler of China. On board is a learned man who speaks Arabic and Hebrew; Columbus thinks those languages will help him talk to the people of Cathay. When the sailors cast off, it is with a feeling of excitement. They know that if they make it, this will be one of the great voyages of all time. They hope to return with gold and spices. Spices make food taste good even if it is a bit spoiled. In these days before refrigerators, spices are very valuable.

The ships stop in the Canary Islands for supplies and perhaps courage; then, on September 6, they head out into the unknown ocean. Columbus has his compass and an astrolabe to guide him. The astrolabe tells him how high the North Star is above the horizon. With

He has many names, too: Cristobal Colón to the Spanish, Cristoforo Colombo to the Italians.

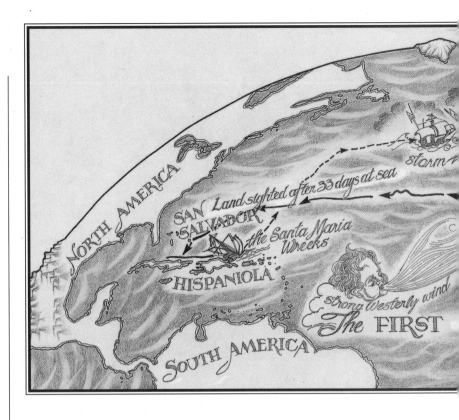

it, he is able to tell his latitude. If he follows a line of latitude, he can keep a straight course. That sounds easier than it is. The rolling of the ship makes the astrolabe readings uneven. Columbus is one of the best sailors the world will ever know. Being a good sailor, he keeps a record of the winds, the speed of the ship, and the compass direction. That tells him how far he has gone in the vast sea. Calculating your position, or longitude, this way is called "dead reckoning." Besides, Columbus seems to have a natural sense of the best way to go: he is famous for being able to find his way at sea. He picks a route with fair winds. Still, it is frightening to go where no one has gone before.

In mid-September they come to what seems to be a meadow of grass in the middle of the ocean. It is the Sargasso Sea—an area of thick, green seaweed. The sailors have never seen anything like this. They are afraid the ships will get tangled in the green muck. But soon they are out of it and into the open sea again. Now there is a fierce storm with waves that rise higher than the church towers in Palos. The ships are sturdy and the seamen skilled, so they survive the tempest. But the sailors are discouraged and fearful. The sea seems endless. On October 9 they say they will go no farther. Columbus pleads for three more days of sailing. Then, he says, if they don't see land they may cut off his head and sail home in peace.

Magnetic North

Magnets always point north, don't they? Well, not quite. Columbus was the first person to discover that. As he sailed west his compass direction changed a bit. Columbus knew something was wrong, but he didn't tell his crew. They would have panicked had they thought the compass couldn't be trusted. Now we know a compass points not to the North Pole but to a magnetic pole nearby.

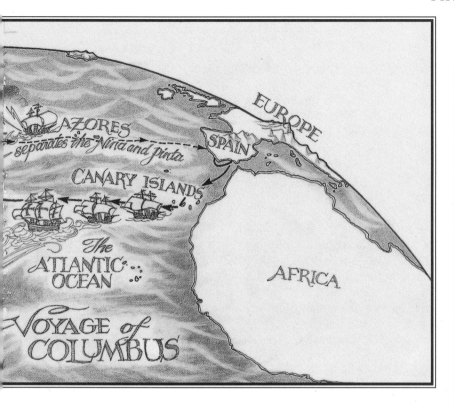

VOYAGE of COLUMBUS

Mutiny—it means revolt or rebellion. Some sailors talked of taking over the ship and heading back to Spain. That would have been a mutiny.

Cabin Boys

Every sailing ship had boys as sailors. Boys were obedient, easy to control, and good at climbing the sails' ropes, or rigging. Every half hour a boy shouted out the time. Here is the call for five o'clock:

Five is past and six
*   floweth;*
More shall follow if God
*   willeth.*
Count and pass,
Make voyage fast.

Three days later, on October 12, a lookout high on the *Pinta*'s mast yells, *"Tierra! Tierra!"* It is the Spanish word for land. The Bahama Islands are straight ahead.

They have made it to the Indies! Columbus was right after all, or so they think. Columbus names the island where they land San Salvador; that means "Holy Saviour." He plants a cross and a Spanish flag on San Salvador. Columbus is a religious man; he believes it is God's wish that he sail and conquer in the name of a Catholic king and queen.

Soon he knows he is not in China. That doesn't bother Columbus. Marco Polo wrote that there were thousands of islands in the Indies. San Salvador must be one of those islands. Japan, China, and the rest of the Indies are sure to be nearby. The island is small but splendid, with tall trees, gorgeous birds, a beautiful beach, and friendly people.

Columbus calls the people Indians. He is puzzled when they don't understand Arabic. Actually, the language they speak is Arawak. They are members of a Taino (TY-no) tribe,

King Ferdinand of Spain watches regally from his side of the Atlantic Ocean as Columbus and his crew are greeted by Arawaks.

Why isn't Columbus looking for land through a telescope?

It hadn't been invented yet.

Historians aren't sure about this, but many believe that the Caribs (in their language it means "valiant people") were cannibals. One thing is sure: the Tainos feared them. That may explain why the Tainos were so eager to help Columbus. Perhaps they wanted him as an ally against their enemies.

although soon others will be calling them Indians. Columbus says of them: "They remained so much our friends that it was a marvel;...they came swimming to the ships' boats, and brought us parrots and cotton thread...and many other things, and in exchange we gave them little glass beads....Finally they exchanged with us everything they had, with good will."

Columbus is lucky. If he had landed on a nearby island, he would have been greeted by the Caribs—a tribe of warriors—and might not have lived to tell the world of his discovery.

The Tainos are peaceful fisherfolk. They welcome the voyagers who have come in bright ships and brought shining beads that seem to capture the sunshine. But what must be in the Indians' minds when they first see these men? Do they think it strange that they wear heavy clothing in a warm land? (Columbus says they wear nothing at all.) Are they surprised that the strangers have skin the color of melons, or that one—a black man from Africa—is dark as chocolate? (Columbus says they are handsome and that their skin is brown, their hair straight.) Do the smells of the seafarers bother them? (The Europeans do not bathe.) Whatever they think, the swords that Columbus and his men carry help convince the natives to do as the strangers wish.

Besides, they want to please. The Taino are generous and intelligent people. Columbus says they learn Spanish words quickly. They also communicate through sign language.

The Tainos don't realize that they do not have long to live. Columbus will kidnap some and take them to Europe as trophies of his voyage. He will help turn many of them into slaves. Soon all the Arawak-speaking tribes will be dead—killed by European weapons, slavery, and diseases. Those diseases—like smallpox and measles—are new in this hemisphere. The natives have no immunity to them.

But that is to come. At first the Tainos help Columbus. He is determined to find gold and the Grand Khan. The Tainos take him to a huge island they call Colba. It is Cuba. Here there are many natives, and some are wearing ornaments of gold! Yet the Grand Khan is nowhere to be found. (Columbus is not discouraged: China and Japan must be nearby.) These Indians seem to be fire-eaters: they put a smoking weed in their mouths. It is the first time the Spaniards have seen tobacco.

There are pearls on Cuba, and enough gold ornaments to take samples to please King Ferdinand and Queen Isabella. Columbus sails home to Spain with brightly colored parrots, Indians, and gold trinkets. Can you imagine what happens when he arrives in Spain? Hardly anyone believed they would ever see him again. But he has found Cathay! At least that is what he says and believes. Now he is a great hero. The king and queen name him Lord Admiral of the Ocean Seas.

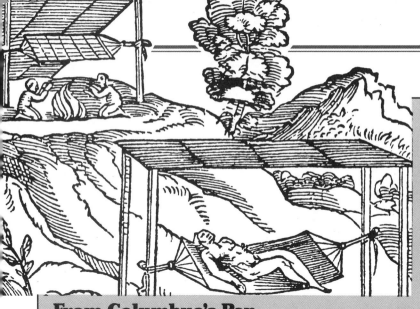

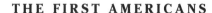

Until Columbus met Indians who slept in them, hammocks, so useful on a ship, were unknown to Europeans.

the natives, that from them we might gain some information of what there was in these parts; and so it was that we immediately understood each other, either by words or signs.

They...believe that I come from heaven...wherever I went ...[they ran] from house to house and to the towns around, crying out, "Come! Come! and see the men from heaven!"

As for monsters, I have found no trace of them except at the point in the second isle as one enters the Indies, which is inhabited by a people considered by all the isles as most ferocious, who eat human flesh. They possess many canoes, with which they overrun all the isles of India, stealing and seizing all they can.

—Christopher Columbus

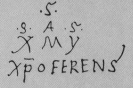

This is Columbus's signature. The bottom word is half Greek, half-Latin: it says "Christ-bearer"—which is what Christopher means in Latin.

From Columbus's Pen

These are the actual words of Christopher Columbus, written in a letter in 1493 to Luís de Santangel, Treasurer of Aragon, Spain, telling of his discovery:

Sir;

...I write this to tell you how in thirty-three days I sailed to the Indies with the fleet that the illustrious King and Queen, our Sovereigns, gave me, where I discovered a great many islands, inhabited by numberless people; and of all I have taken possession for their Highnesses by proclamation and display of the Royal Standard....

[Hispaniola is] full of trees of endless varieties, so high that they seem to touch the sky, and I have been told that they never lose their foliage. I saw them as green and lovely as trees are in Spain in the month of May. Some of them were covered with blossoms, some with fruit....There were palm trees of six or eight varieties....There are wonderful pinewoods, and very extensive ranges of meadowland. There is honey, and there are many kinds of birds, and a great variety of fruits....Hispaniola is a marvel.

...[The Indians] are well-made men of commanding stature, they appear extraordinarily timid. The only arms they have are sticks of cane, cut when in seed, with a sharpened stick at the end, and they are afraid to use these. Often I have sent two or three men ashore to some town to converse with them, and the natives came out in great numbers, and as soon as they saw our men arrive, fled without a moment's delay although I protected them from all injury.

...they are so unsuspicious and so generous with what they possess, that no one who had not seen it would believe it.

...in the first isle I discovered, I took by force some of

17 The Next Voyage

The pineapple was just one of many foods new to the Spanish.

If you're going to be an explorer, you need a base: a place you can go for supplies and help. Columbus knew that, so on his first voyage he set up a base on the island of Hispaniola (the island is now divided down the middle into two countries, which we call Haiti and the Dominican Republic). He thought the base would become an important trading post when he found the Grand Khan.

Hispaniola was the first Spanish settlement in the Americas—and it flopped. As soon as Columbus sailed back to Spain for more ships and men, the settlers he left behind started fighting over gold and Indian women. Soon they were killing each other. The Indians—who must have been angry at the way they were being pushed around—killed most of the rest of them.

While this was going on, Columbus was in Spain being a hero. Now Isabella and Ferdinand were happy to give him ships and men. After all, he had found the Indies, he was sure. "It's just a matter of getting past those outlying islands to reach Cathay," he must have said.

His second trip was to be the payoff voyage. Now that he knew the way, it wouldn't be difficult to cross the ocean. This time he had 17 ships and 1,200 men. He took horses and armor and European goods. Everyone was sure Columbus would meet the Grand Khan and come home with boats full of gold and silk and spices. So adventurers from some of the most important families in Spain went with him.

The adventurers were nothing but trouble. They expected to find China, and when they didn't find it they blamed Columbus.

To keep the men happy, Columbus gave them land on the islands he discovered. They soon began capturing Indians and using them as slaves. Then Columbus sent a boatload of Indians back to Spain to be

The Europeans called America a "new world"—but it was another old world with its own ancient civilizations and peoples. They were just different from those in Europe.

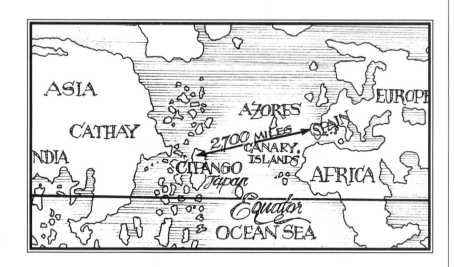

ASIA

CATHAY

INDIA

AZORES

SPAIN

2700 MILES

CANARY ISLANDS

CIPANGO
Japan

AFRICA

Equator

OCEAN SEA

EUROPE

Cipango = Japan
Cathay = China

sold as slaves. It was a poor way to begin in a new land—especially as the Indians never made good slaves. They just died off.

Columbus kept searching for gold mines, but he didn't find any. He never guessed that the Caribbean Islands would make some Europeans very rich—but with sugar, cotton, and tobacco, not gold. Huge plantations would produce crops for Europe's markets, creating enormous wealth.

Because workers are needed to grow crops, and because Spaniards didn't want to work in the fields—and the Indians were dying—black people would be brought from Africa to be fieldworkers. The first Africans would come in 1503; by 1574 there were 12,000 black

Top: Columbus's idea of the length of his voyage and the whereabouts of his hoped-for destination. Below: what was really there—and how far it was.

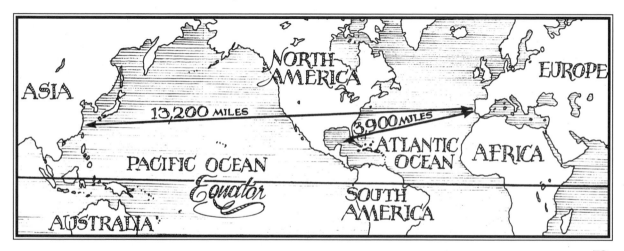

ASIA

NORTH AMERICA

EUROPE

13,200 MILES

3900 MILES

PACIFIC OCEAN

ATLANTIC OCEAN

AFRICA

Equator

SOUTH AMERICA

AUSTRALIA

These Indians are preparing a corn liquor drink. The woman in the foreground isn't throwing up; she's mixing the corn with saliva to start fermentation. Corn wasn't the only new food Europeans found in America. More are listed below.

The Columbian Exchange

From the Old World to the New World:

Horses, cattle, pigs, sheep, chickens, honey bees, wheat, Asian rice, barley, oats, soybeans, sugarcane, onions, lettuce, okra, peaches, pears, watermelon, citrus fruit, rye, bananas, olives, chickpeas

From the New World to the Old World:

Corn, potatoes, tomatoes, peppers, chocolate, vanilla, tobacco, beans, pumpkin, cassava, avocado, peanuts, cashews, pineapple, blueberries, sunflowers, squashes, quinine, wild rice, turkey, marigolds, petunias, sweet potatoes

pers, beans, pumpkins, and tomatoes were growing in Europe. One plant—the potato—proved more valuable to the Old World than all the gold in both the Americas.

The exchange went both ways. Columbus brought oranges to the West Indies, where they were unknown. Cattle, sheep, horses, and pigs were sent to America from Europe. That transfer of plants and animals provided a way to feed bigger and bigger world populations. Corn was soon a basic food in Africa, as were potatoes in Ireland. (Sometimes this reliance on one kind of food became what is known as a mixed blessing. In the 19th century, for instance, the potato crop in Ireland failed several years running. Thousands of poor Irish people died of starvation.) The American sweet potato became important in China. Italians took the tomato and created a new sauce for spaghetti.

But nobody in Spain was interested in agricultural revolutions. It was gold and spices that they wanted. So Columbus was no hero when he returned from his second voyage. He had a few gold pieces,

Africans on Hispaniola —and that was just one island. It was the beginning of black slavery in America.

Columbus did something important that no one noticed at first: he helped start an agricultural revolution. Before long, American corn, pep-

What If Columbus

Suppose the old geographers had been right. What if the world were tiny and there were no American continents? What if Columbus had got to Cathay? Would he have met the Grand Khan? Would he have returned home in ships laden with gold and spices?

We can be sure about the answer to only one of those questions. We know Columbus would *not* have met the Grand Khan. There was no Grand Khan.

Most of what Columbus knew of Cathay came from Marco Polo's book, and it was old knowledge. The Grand Khan and his family had been replaced in the 14th century by rulers of the Ming family. The word *khan* was no longer used in China. Columbus even had China and India confused.

In Columbus's time, at the end of the 15th century, China was the most advanced civilization in the world. Its people were better fed, better housed, better clothed, and better educated than anyone on

but he hadn't found a source of gold, and he hadn't found China.

Still, if Columbus had stopped then, he could have retired with wealth and honors. But he was the kind of man who never stopped. That's what made him a great explorer: he kept going. He made two more voyages—four trips in all. He discovered more islands and the mainland of South America. He never did discover North America, and he was always convinced he had found the Orient. He was sure that all those islands he found were off the coast of Cathay. South America bothered him because it did seem like a mainland. It had a great river. He knew it wasn't China; China couldn't be that far south.

Columbus solved the problem of the southern land by deciding it was the Garden of Eden, the place

King John of Portugal said Columbus was "a big talker and boastful in his accomplishments." Can you think why bragging might sometimes be useful and not just conceited?

Had Reached Cathay?

the globe. Most Chinese lived in family-centered farming villages. They were led by a hereditary emperor, but government officials—often scholar-artists chosen by fair examinations—ruled the nation. Anyone, from any village, had a chance at success.

Chinese technology was way ahead of that of the West. The Chinese had been printing books for centuries. Chinese mathematics, astronomy, ship design, and navigation aids were more sophisticated than anything the Spanish or Portuguese had developed. The great admiral Zheng He, a Chinese Muslim, sailed from China across the Indian Ocean to Africa 60 years before Columbus's voyage. Zheng He commanded a fleet of more than 300 ships, some three times longer than the *Santa María*. Zheng He called on foreign rulers, brought lavish gifts, and encouraged tribute to China, but he never tried to have China rule other nations.

Chinese sailed to Africa in junks like these, years before Columbus ever made it to America.

Why, with their advanced culture, didn't the Chinese lead the world in exploration? Why, after the 15th century, did Western nations begin to dominate the world instead?

Perhaps the Chinese were so self-sufficient that they didn't think they could learn from others. Perhaps they were too smug, too content with themselves.

Competition drove the European nations; the Chinese didn't feel a need to compete with anyone.

If Columbus had made it to Cathay, he would have been treated with courtesy but probably not allowed to see the emperor. The Chinese would have thought him a barbarian. Columbus, with his tiny fleet and his scruffy sailors, would not have impressed them.

the Bible describes as the first home of Adam and Eve. Columbus believed the Garden of Eden must be somewhere on earth. When he saw brilliantly colored birds and flowers in the southern land, he was sure he had found it. South America was like an Eden. He wasn't the only explorer to be fooled.

Living at the Center of the Universe

When, in the 20th century, men reached the moon, it was an astonishing feat of technology. People actually built a spaceship and rocketed it away from the pull of earth's gravity and on through space. But even so, everyone knew the moon was there. They even knew in advance what the moon would look like when they got there. That wasn't the case in 1492. No one was prepared for Columbus's discovery.

Imagine that you live in the 15th century. Put yourself in school in Egypt or Korea or Italy. Your lesson for today is about world geography. The teacher is telling you that the earth is the center of the universe. But since you're smart, you already know that. Everyone in the 15th century knows that. You also know that the earth doesn't move. The sun and stars move around the earth, the teacher says.

Today you will learn that this world was created with one great three-part mass of land (divided into Africa, Asia, and Europe) and one great ocean that laps its shores. The seas and rivers and lakes are like highways to be traveled. The great ocean is no highway; it is a moat that protects the land. The world is orderly and easily understood.

So when news comes of the voyage of Columbus, it never occurs to you to think that he has found a new continent. Columbus himself couldn't believe what his eyes told him; why should you consider it? And when, in the early 16th century, it finally becomes clear—Europe, Asia, and Africa are not the only continents—you realize that some things you were taught in books and school and church were wrong. The Old World is not the center of the universe and not as easy to understand as you once believed. It makes you begin to ask new questions. It makes you begin to think for yourself.

The same kind of thing happened to Indian girls and boys in America. They, too, thought they lived at the center of the

This map was a bit out of date even in 1436, when it was made. But most people thought the world looked roughly like this. Turn the book sideways to help you find Spain and Europe.

universe. They knew of no other continents or cultures—until some brightly painted ships came from the east and changed their view of the world forever.

18 Stowaways: Worms and a Dog

Columbus used this almanac's prediction of an eclipse of the moon to fool the Indians into helping him.

On Columbus's fourth voyage he was attacked. Who was the enemy?

Worms. Yes, you read that right, the enemy was worms—worms that ate holes in the ships' bottoms. Worms did so much damage that the ships were ruined and couldn't sail anymore. Columbus and his crew were marooned—stuck—on the island of Jamaica.

Now if you're going to get marooned, Jamaica is a pretty nice place to be. It's a beautiful island where foods grow easily. However, the Spanish explorers (and later the English explorers) weren't interested in growing their own food. Columbus and his men sat around waiting for the Indians to feed them. At first the Indians did that, but then they said something like "Get your own food." The explorers didn't know how, and they began to starve. Then the Indians decided to attack.

The situation was serious when Columbus got out his scientific papers and read something that gave him an idea.

Columbus had Abraham Zacuto's charts of the stars—astronomical charts. Zacuto was a leading scientist and a Portuguese royal mathematician. His charts helped sailors figure out latitude. They also told about eclipses of the moon and sun. Columbus read in Zacuto's charts that there would be an eclipse of the moon on the last day of February in 1504.

"They all made fun of my plan then; now even tailors wish to discover," wrote Columbus about the way people changed their ideas after he returned from his first voyage.

77

Balboa made it across the jungles and mountains of Panama without losing any men. His stowaway dog Leoncico made it, too; that's him frisking at the edge as his master strides into the Pacific Ocean.

Balboa and other Spanish explorers were ***conquistadors*** (kon-KEES-tah-dors). It is a Spanish word that means "conqueror."

So on that day Columbus called the Indian chiefs together and told them he had power over the moon. He said if they didn't give his men food, he would blot out the moon. It happened just as Columbus said it would, and the Indians cried and begged Columbus to bring back the moon. He agreed, the eclipse ended on schedule, and his men never went hungry again. (You can see Zacuto's tables, with Columbus's notes, if you go to Seville, Spain.)

Finally, after more than a year, Spanish ships rescued Columbus and his men. Columbus went home to Spain, but now no one paid any attention to him. Isabella was dead and Ferdinand wouldn't even see him. Most people thought the islands he had discovered were worthless. It was only after Columbus died that people began to realize the value of his discoveries.

The most important thing he did was to sail into the unknown. That took great courage and skill. Once he showed it could be done, others followed. Among them was Giovanni Caboto (jo-VAH-nee kah-BOW-tow), an Italian who went to England, where he was called John Cabot and given a small ship. Cabot sailed across the Atlantic in 1497 with only 18 sailors. He had to be very brave to do that. He landed in Newfoundland, where the Vikings first landed 500 years earlier. Later the English claimed all of North America because of Cabot's voyage.

If you think exploring is just a matter of luck, consider the story of Vasco Nuñez de Balboa (VASS-ko NOON-yez day bahl-BOW-ah), one of the greatest of the Spanish explorers. He was a stowaway, which means he hid on a ship. The ship was heading for Darien (now called Panama) to search for gold.

Balboa hid in a flour barrel with his dog, Leoncico (lay-on-SEE-ko). He waited until the ship was far out at sea before popping out of the barrel. Why was he hiding? Balboa owed money to some people on the island of Hispaniola, and he couldn't pay his debts.

So if you'd asked his creditors (the people he owed money to), they would have told you that Balboa wasn't a very nice person. But if you'd asked others who knew him, they would have said something else. Balboa was a born leader. By the time the ship landed in Darien, in 1513, he was in command. He sent the incompetent leader, Enciso (en-SEE-zo), home in chains. (That was a mistake—you'll see why.)

Then Balboa established the first permanent European settlement in the Americas. That would have been enough to get him into the history books, but it wasn't enough for Balboa. He wanted to find gold. It wasn't gold that made him famous, though. Balboa was the first European to see the Pacific Ocean from the American continent. He "discovered" the Pacific for the peoples of Europe. An Indian chief, Comaco, told Balboa about that ocean: "When you cross over these mountains you shall see another sea, where they sail with ships as big as yours, using both sails and oars as you do, even though the men are like us." (Before Balboa's discovery, Europeans thought there was only one ocean. They called it the Ocean Sea.)

Balboa married Comaco's daughter, so the Indian chief helped the white men even though he was disgusted by their constant fights over gold. Here are more of Comaco's words, written down by Peter Martyr, who lived in those times: "What is the matter, you Christian men, that you so greatly value so little gold more than your own peace of mind?" (What did he mean by that? Do you think peace of mind was more important to the Indians than gold? Were they wise or foolish?)

"I will show you a region flowing with gold, where you may satisfy your appetites," said Comaco. Of course Balboa was anxious to go for the gold, but he also wanted to know about the sea that Comaco described. He had an explorer's curiosity.

Balboa decided to march across Panama. Let's go with him. Getting to the Pacific will not be an easy jaunt. Imagine smothering heat, pounding rainstorms, and jungles so thick you can hardly hack your way through them. Add killer bugs, snakes, and germs—and you'll begin to get the picture. Balboa and his men are wearing padded leather jackets and, on top of that, metal armor. Can you see them in the jungle in those hot, heavy garments? They make it to the Pacific Ocean and back—and not a man dies. It is quite an accomplishment.

If you are an explorer, discovering is not enough. You also have to tell people what you find. Balboa did that, too. He sent news of the Pacific Ocean to the people in Spain. Then he started organizing his next exploration. He was planning to go south, to the region flowing with gold. It was called Peru.

But in the meantime, back in Spain, Enciso was thirsting for revenge. He accused Balboa—falsely—of treason. People in Spain believed Enciso. A new governor was sent to Darien with orders to get rid of Balboa. He did. Balboa was beheaded; his head was stuck on a pike for all to see. The officer who arrested Balboa was named Francisco Pizarro (pih-SAR-oh)—remember that name.

When Columbus arrived in San Salvador in 1492, he set off a cultural tornado. He, and those that followed, brought ideas, technology, and germs that overwhelmed the lands they invaded. In that way, they did make a New World.

Balboa was brave and adventurous, but of course he wasn't the first European to see the Pacific—he was just the first European to see the Pacific's western coast. Explorers and traders knew the eastern side quite well.

19 Sailing Around the World

Magellan had been to the Spice Islands, sailing around the Cape of Good Hope and across the Indian Ocean. Now he wanted to find the way west.

Exploring can be a dangerous business. It certainly was dangerous for Ferdinand Magellan (muh-JELL-un). His voyage was perhaps the most remarkable of all. Magellan was the explorer who actually found China by sailing west from Spain. He discovered a passageway—a strait—near the tip of South America, sailed through it, and went on across the Pacific Ocean. That southern passageway is so treacherous and stormy that even now only the most skilled sailors attempt it. It may be called a strait, but it is crooked, with steep, rocky walls. It took Magellan 38 days to get through the strait that was later named for him. But the worst was yet to come. No one guessed that the Pacific was as huge as it is. Magellan headed on, right across that ocean. His expedition would make it around the world.

But that wasn't what he intended when he left Seville, Spain, in 1519, with five ships and about 270 men. He was heading for the Spice Islands. The Spice Islands are also known as the Moluccas (muh-LUH-kuz). If you look at a map, you'll see them in Indonesia, just south of the Philippine Islands and west of New Guinea.

Europeans had been to the Spice Islands by heading south, going around the tip of Africa, and then sailing east. Magellan was convinced he would find a shortcut if he went in the other direction—west— away from Africa and across the Pacific. He thought the Pacific was a calm ocean, much smaller than the Atlantic. (Remember Ptolemy's mistake? Magellan had studied Ptolemy, too.) Magellan's original plan was to go to the Spice Islands by the new "short" route, across the Pacific, and then turn around and come back the same way. (The best maps of the day showed Japan a few hundred miles west of Mexico.) Magellan was in for a big surprise.

Right from the beginning the voyagers had adventures. When they first reached South America they met a tribe of cannibals. What would you do if you met cannibals? Magellan and his men sailed away quickly. They went down the unknown South American coast, landed again, and finally made camp for the winter. Here is how one of them described what happened next:

> One day [without anyone expecting it] we saw a giant who was on the shore, quite naked, and who danced, leaped, and sang, and while he sang he threw sand and dust on his head. Our captain sent one of his men toward him, asking him to leap and sing like the other in order to reassure him and show him friendship. Which he did.
>
> Immediately the man of the ship, dancing, led this giant to a small island, where the captain awaited him. And when he was in front of us, he began to marvel and to be afraid, and he raised one finger upward, believing that we came from heaven. And he was so tall that the tallest of us came up only to his waist....He had a very large face, painted round with red, and his eyes also were painted round with yellow, and in the middle of his cheeks he had two hearts painted. He had hardly any hairs on his head, and they were painted white.
>
> The captain caused the giant to be given food and drink, then he showed him other things, among them a steel mirror. Wherein the giant seeing himself was greatly terrified, leaping back so that he threw four of our men to the ground....The captain named the people of this sort Patagoni.

Patagones is what the Spanish actually called them. Magellan and his men stayed five months with the Patagonians. When they left, they took two of the giants with them—in chains. They wanted to display them when they returned to Spain.

Now the vast Pacific was ahead of them. Had they known what they would encounter, they might have turned back. Look at a map of the Pacific Ocean. Now imagine yourself in a small ship, perhaps 70 feet long, heading west from the tip of South America and not knowing where you are going.

Pacific means peaceful or peace-loving. Naming an ocean "pacific" is what's known as wishful thinking.

Patagon was a dog-headed monster in a 16th-century Spanish novel that Magellan may have read. The South American natives wore bushy furs and had painted faces; maybe, at first, they seemed like monsters. Magellan's crew thought the Patagonians' feet were especially big. But what really made their feet look big were their roomy moccasins, stuffed with straw to keep out the cold! (Don't forget, the tip of South America is only 500 miles from the Antarctic Circle.)

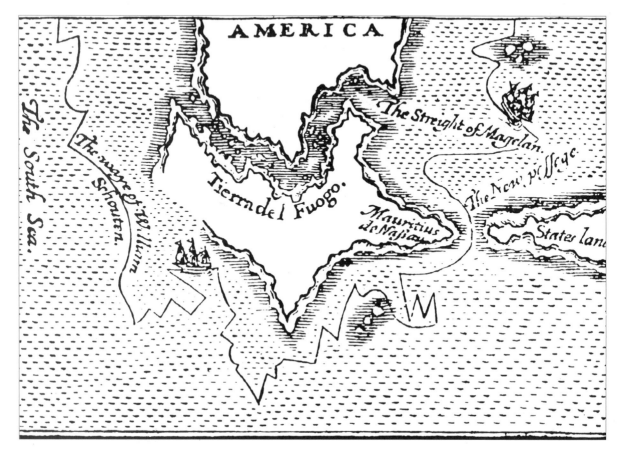

This map of the Strait of Magellan was made 100 years after Magellan's voyage. By then people knew what Magellan did not: that Tierra del Fuego, the "land of fire" at the tip of South America, was an island and that you could sail around it from the Atlantic to the Pacific. Magellan thought the Land of Fire was the edge of a huge Antarctic continent.

Patrid (PEW-trid) means disgusting or rotten.

Magellan happened to be a man of vision. When he realized the vastness of the ocean he was crossing, he didn't change his plans. He decided to go for it. He decided to sail around the world.

Like it or not, his crew was stuck with his decision. They were men from nine different nations. One was an Englishman, and another was a slave captured in the Moluccan Islands (who was expected to be an interpreter). One crew member kept a journal.

He was Antonio Pigafetta, and some say he was a spy from Venice. We wouldn't know the details of Magellan's voyage if it weren't for his written record. One of the things Pigafetta said in his journal was that no one should ever attempt the voyage again. It was just too dangerous. What he didn't realize is that it is the unknown that is most frightening. The knowledge he brought back allowed people to sail off with confidence.

But for Magellan and his men it was a terrible voyage. The course they steered missed every island between South America and Guam. By the time they reached land, near China and the Spice Islands, most of them were almost dead of hunger. Many did die. They ate rats and chewed leather straps and drank putrid water. Pigafetta wrote:

We were three months and twenty days without getting any kind of fresh food. We ate biscuit, which was no longer biscuit, but powder of biscuit swarming with worms, for they had eaten the good. It stank strongly of the urine of rats…And of the rats…some of us could not get enough. . .

Finally, after a battle with angry natives on an island that may have been Guam, they landed in the Philippine Islands. Enrique, the Moluccan slave, spoke to the islanders. The Filipinos understood him. Magellan realized he had crossed the Pacific! (Enrique had been captured by Spanish traders and taken from the Moluccas around Africa to Spain. He was the first person to sail around the world.)

In the Philippines the famished sailors found fresh water, food, and the King of Cebu. Magellan told him about the Spanish king and said he was the greatest in the world. Then Magellan told the king of his God and converted him to Christianity.

"The king said he was content," wrote Pigafetta, "and that if the captain wished to be his friend, as a greater token of love he would send him a little of his blood, from the right arm, and that the captain should do likewise."

Magellan stayed in the Philippines, had many adventures, and learned the ways of the people. He spoke kindly to the islanders, and many became Christians because of him. Then he faced a problem other explorers would face: how do you deal with your new friends' enemies?

Magellan chose to fight them. It was the wrong choice. Pigafetta described what happened when they joined the Philippine Islanders' war against a chieftain named Lapu Lapu:

[The enemy forces] followed us, hurling poisoned arrows four or six times; while, recognizing the captain [Magellan], they turned towards him…[and] hurled arrows very close to his head. But as a good captain and a knight he still stood fast with some others, fighting thus for more than an hour. And as he refused to go back, an Indian threw a bamboo lance in his face, and the captain immediately killed him with his lance….Then, trying to lay hand on his sword, he could draw it out only halfway, because of a wound from a bamboo lance that he had in his arm. Which seeing, all those people threw themselves on him, and one with a large javelin…thrust it into his left leg, whereby he fell face downward. On this, all at once rushed upon him with lances of iron and bamboo and with javelins, so that they slew our mirror…our comfort, and our true

Magellan

Magellan was born into a noble Portuguese family. When he was a boy, he served as a page in the royal household. Then he went to India, where Portugal was powerful. But when he asked the Portuguese ruler to let him sail west from Europe to the Spice Islands, he was turned down. He went to Spain instead. He set out from Seville and headed into the Atlantic at Sanlucar; today it's a seaside resort.

83

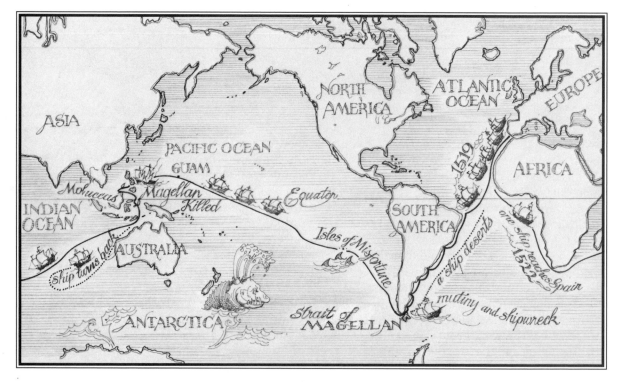

Have you ever gotten lost and not known where you were walking to? Think of that feeling, and then think of Magellan and his men, sailing on over the endless ocean with no sight of land until they reached the island of Guam.

guide....Then, seeing him dead, as best we could we rescued the wounded men and put them in the boats which were already leaving.

Those 15th and 16th century explorers were brave and determined. Few men could have accomplished what Magellan did. In his journal Pigafetta said he hoped future generations would not forget his captain:

For among his virtues he was more constant in a very high hazard and great affair than ever was any other. He endured hunger better than all the others. He was a navigator and made sea charts. And that that is true was seen openly, for no other had so much natural talent, boldness, or knowledge to sail once around the world, as he had already planned.

Finally, Magellan's voyagers reached Spain—18 men on one battered ship. (How many men and ships were there when Magellan started out?) They had been gone almost three years and had circled the globe. Imagine a ship from a distant galaxy landing on earth today. That was how amazed people were in Spain when Magellan's ship returned. The voyagers brought news of an unknown world—it just happened to be their own world.

20 What's in a Name?

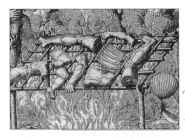

Vespucci said he saw people in the New World who did strange things—like eating other people.

Your teachers probably tell you that writing is important. Well, if ever there was a story to prove that, it's the story of Amerigo Vespucci (vess-POO-chee).

Columbus did all that discovering, and yet it was Amerigo whose name got put on the land. History isn't always fair.

Not that Amerigo wasn't a good guy. He was. And quite an adventurer, too. He made several trips across the ocean and went way down the coast of South America.

But what made him famous was that he wrote about his trips. His letters were so fascinating that everyone wanted to read them. And, thanks to Gutenberg's printing press, many people did. Amerigo Vespucci did something else that was important. He understood that there was a huge continent over here, a continent new to Europeans, and he said so in his writings. He called it a New World. Actually, he called it a *novus mundus* because, like educated people of his time, he wrote in Latin. (His name in Latin was Americus Vespucius.)

Columbus and Vespucci were friends, but their personalities were different. Columbus was a dreamer; Vespucci was a hardheaded businessman whose hobbies were mathematics, sailing, and reading. Columbus couldn't give up his dream of finding Cathay; Vespucci looked at the land of South America with an open mind. "In those southern parts," he wrote, "I have found a continent more densely peopled and abounding in animals than our Europe or Asia or Africa." He could see that it was a vast continent and that it wasn't China.

A man named Martin Waldseemüller (valt-ZAY-mew-ler) actually named America. He was a printer and mapmaker who was fascinated

One famous historian had this to say about how America got named: "America was discovered accidentally...and most of the exploration for the next fifty years was done in the hope of getting through or around it. America was named after a man who discovered no part of the New World. History is like that, very chancy."

On one of his trips to South America, Vespucci and his companions became the first Europeans to lay eyes on the Amazon River. On his second trip he and two other explorers reached a huge bay and another river. They named it Rio de Janeiro, because they found it in the month of January.

by Vespucius's letters. He said, "Americus Vespucius has first related without exaggeration of a people living toward the south, almost under the antarctic pole. [They] go around entirely naked, and not only offer to their king the heads of their enemies whom they have killed, but also feed eagerly on the flesh of their conquered foes."

You can see why this made interesting reading. In 1507 Waldsee-müller printed a huge world map, and on it he put a new continent. He used information from the voyages of Columbus, Cabot, and Vespucci to make the map. He decided to put the name AMERICA on the southern continent, since Amerigo had written about it.

His map was the most up-to-date and reliable one around. Many people bought it. Six years later Waldseemüller made a new map. This time he didn't think it was such a good idea to call the new land America. He left that name off, but it was too late. People were already using the name, and it stuck. Amazing, isn't it, how fickle history can be?

In one corner of his new world map, Waldseemüller included a portrait of the man whose first name he'd attached to the great continent across the sea—*America.*

21 About Beliefs and Ideas

Human sacrifice was a fundamental part of the religion of the Aztec people, who lived in what is now Mexico.

If you want to believe in great green dragons, go ahead; no one will stop you. In the United States we are free to believe anything we want.

If you believe in a moon god, you can even start a Moon God religion. Some people may think you are a bit weird, but no one will throw you in jail.

It wasn't always that way. There was no religious freedom in America at first. The very idea would have seemed strange to the Native Americans or to Columbus and Queen Isabella. In both America and Europe religion and government were bound together. This was expected, but it sometimes led to disaster. The Aztec Indians killed tens of thousands of people who held beliefs different from theirs. Europeans spent centuries fighting religious wars before some of them began to question whether it was right to force others to believe as they did.

In 15th-century Europe most people were Roman Catholic. There were some Jews and Muslims, too, but no Protestants. (The Protestant churches hadn't been founded yet.) Roman Catholicism and Eastern Orthodox Catholicism were the only Christian religions.

The Protestant religions would get started in the 16th century when a man in Germany named Martin Luther protested and tried to reform the Catholic church. There were 95 things in the Catholic

Roman Catholicism is centered in Rome, Italy, and is led by the Pope. Eastern Orthodox Catholicism is divided into regional churches led by *patriarchs.* The leading Orthodox church is in Istanbul (formerly called Constantinople), Turkey.

A *reformer* is someone who wants to change the world and make it better.

Speaking Out

History teaches us that there are always a few clear-sighted people who resist popular opinion and stand up for what they think is right, even if that means that most people laugh at them or send them to jail. Some people in the 15th and 16th centuries spoke out against slavery. Some spoke out against the expulsion and murder of Jews and Muslims in Spain. Some spoke out against torture. But most people didn't listen to them.

church that Luther thought should be changed. He wrote them all down and nailed the list to a church door. Some people agreed with Luther's list; many didn't. Those who agreed were called "protesters." They started new Christian religions: *Protest-ant* religions. Because the protesters thought of themselves as reformers, the time they lived in is often called the Reformation.

Unfortunately, the clash between Protestants and Catholics led to centuries of hate and violence in Europe. Instead of talking calmly about their differences, Protestants and Catholics fought about them. There were terrible wars that split towns and families. Neighbors and relatives killed each other because they thought differently about religion, and yet all of them claimed to be Christian. Many people went to the New World to escape from those wars of religion. Many went in search of freedom of belief.

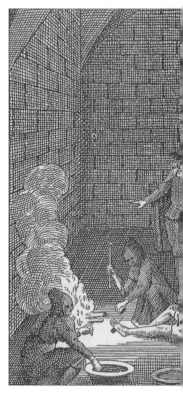

Remember 1492? That was the year Columbus made his first voyage to the New World. It was also the year all Jews had to leave Spain. Those who stayed were forced to convert to Christianity. If they only pretended to convert—and were caught secretly practicing their religion—they were burned alive.

Slavery was common everywhere then, and it didn't seem wrong to many people: not to the Portuguese or the Arabs or the Dutch or the Spaniards or the Africans—who were all involved in selling human beings as slaves.

Some explorers thought they could serve God by converting the Indians to Christianity and, if the Indians wouldn't convert, by killing them. In America the Aztec Indians practiced blood sacrifices; the Iroquois tortured their captives; the Mound Builders kept slaves. They didn't think that was wrong either.

How could people behave that way? Were they different from us?

Not really. Slavery, torture, and religious intolerance have been around for a long time. It is always easy to do and think as everyone else does. And here we are, at one of the most important reasons for studying history: to learn from the mistakes of others.

Columbus had many good qualities, and yet he made slaves of the

Queen Isabella asked a special religious court, called the Inquisition, to come to Spain. They forced people who weren't Catholic to convert or leave the country. If they refused, the court tortured and executed them. As you can see, the court had some imaginative ways of doing this.

Indians. You will read about many good people in history who did terrible things to other people. Usually, they believed they were doing good. They meant to do good. Many people told them that they really were doing good. Does that excuse them? Does it make a difference to the victim? Is it right to force others to think as you do? Is it possible?

Columbus and most Spaniards believed their religion was the only true religion. Suppose you are convinced that your way of thinking is good for everyone. Suppose you are sure you are right. Can you be sure?

Are you bad if you do something wrong but you don't know any better?

These are the questions that philosophers ask. They have no easy answers. Think about them. Talk about them. Write about them.

Many of us are Americans because someone in our family wanted to be free—free to believe whatever he or she wished to believe. This country gave birth to a form of religious freedom unknown anywhere else in the world.

The United States was first, among all nations, to allow people to belong to any church, or to none. But it didn't happen in the beginning, as you will see.

Stories of How the World Began

How did the world begin? Scientists say there was a big bang and dust and particles of matter came together to form stars and planets.

In the Western nations the Bible told of Adam and Eve and how God created the world in seven days.

The Popol Vuh, the sacred book of the Mayan Indians, told this tale of creation:

"This is the story of how all was still, all quiet, in silence; there was no movement, no sound, and the whole of the sky was empty. This is the first story, the first speech. There were no people yet, nor animals, nor birds, nor fish, nor crabs, nor trees, nor stones, nor caves, nor ravines, nor grass, nor forests. Only the sky was."

Into the silence came the Creator and the Maker and they made the dawn and earth and animals. They took clay and made men and women. But the clay people were soft and fell down; they could not stand. So the Creator and the Maker made people of wood and the wood people stood, but they had no minds or souls.

"A flood came from the sky, a great flood that fell on the heads of the wooden dolls."

Then the Creator and the Maker made men and women of straw. But the straw people could not think and the stones and animals rose up against them. The straw people were terrified and turned into monkeys. Finally the Creator and the Maker made men and women from corn, and real people—the Mayan people—came forth.

Among the American nations there were many stories of the creation and many religions. Some of the stories told of ancient times when the animals talked to people. There were stories of Turkey, who was good and helpful; and of Coyote, the mean trickster. Iroquois stories told of Great Turtle, who carried the earth on his back.

The Indian religions were as varied as the peoples. Most Indians believed there were many gods. Some of their religions demanded sacrifice to those gods, even human sacrifice. But most did not. Most included the idea that the gods could be found in all of nature. Because of that, many Indians (but not all; some did not care) tried to live in harmony with the earth, its plants and birds and animals.

To the Maya, corn was the foundation of life. Their sacred corn gods wore corn ears on their heads.

Another Story...

"Before the High and Far-Off Times, O my Best Beloved, came the Time of the Very Beginnings; and that was in the days when the Eldest Magician was getting Things ready. First he got the Earth ready; then he got the Sea ready; and then he told all the Animals that they could come out and play."

FROM "THE CRAB THAT PLAYED WITH THE SEA" IN *JUST SO STORIES* BY RUDYARD KIPLING

22 New Spain

Legend told the Aztecs to build their city on the spot where an eagle perched on a cactus.

In Europe the year was 1519. In the New World the Aztec Indians of Mexico had their own calendar. It was even more exact than the calendar in use in Europe at the time. The Aztec calendar predicted big events during this year.

The Mexican prophets had said that Quetzalcoatl (ket-zal-KOH-a-tul) would come from the east—from the rising sun—to take back the land that was his. Quetzalcoatl, the supreme god, was the god of peace. The prophets had predicted the exact day he would come. So when the explorer from Spain, Hernando Cortés, came on that very day, many believed he was the god whose coming was foretold. Hernando Cortés, tall and regal, was carried on a ship with sails as bright as birds' wings. Strange animals came with him, and men sat on their backs. Never had anyone seen a man on horseback. Some thought that horse and rider, both wearing glistening armor, were one creature—a splendid man-horse, a shining centaur.

Moctezuma, ruler of the great Aztec kingdom, was a thoughtful man. But sometimes he had a hard time making up his mind. Perhaps that was because he didn't have to do much for himself; he had hundreds of servants. Moctezuma was in his favorite palace—it had 100 rooms, 100 baths, walls of marble and rare stone, and courtyards filled with singing birds, flowers, and fountains—when messengers who had run from the coast told him astonishing news: a small army of men and animals, like no men or animals seen before, was now standing on Aztec soil.

Moctezuma II was the last great Aztec ruler. His name meant "angry lord." The Spaniards couldn't pronounce it properly, so they called him "Montezuma."

To the ancient Greeks, a *centaur* was a mythological creature, half-man, half-horse.

The Templo Mayor (Spanish for "great temple") stood in a sacred area in the center of Tenochtitlán. At the top were twin shrines, dedicated to the two most powerful Aztec gods.

At first, Moctezuma was sure it was the great god Quetzalcoatl, so he sent gifts of gold and precious jewels. The next day he wondered: perhaps they weren't gods. Reports from the scouts made the strangers seem like men, so he threatened them. Were they gods? Were they men? He hesitated. That was his mistake.

Cortés was not a man to hesitate; he was a man of action. Sometimes he compared himself to Alexander the Great, the mightiest warrior of the ancient world. There was no modesty in the comparison, but much truth. Cortés became one of the world's greatest conquerors. Moctezuma told him to stay on the coast. Cortés marched toward Moctezuma's capital. When some of his men were fearful and wanted to turn back, Cortés burned his own ships. Now there was no way to go but forward.

He marched through a countryside filled with people and villages and cities. An Indian woman marched with him. She had learned Spanish, become a Christian, and taken the name Doña (DON-ya—it means "lady") Marina. "She was a princess...as her appearance and bearing clearly showed," wrote a soldier who was with Cortés. Doña Marina could talk to the Indians, and so, through her, could Cortés. "The help of Doña Marina was of the highest significance to us," the soldier added.

To *covet* (KUV-it) means to want something badly.

Cortés was heading for the greatest city in the Americas; the Indians called it Tenochtitlán (tuh-nock-tit-LAN). As he marched, messengers came from Moctezuma. The messengers brought gifts of gold, robes of parrot feathers, embroidered cotton cloth, food— and orders. "Do not come to Tenochtitlán. Turn back. We will give you gold. Turn back," ordered Moctezuma. Cortés marched on.

The Aztecs watched from canoes as Cortés's army entered Tenochtitlán along a causeway. The city was built on a lake protecting it like a moat.

Cortés was helped by an Indian woman, Doña Marina. She learned Spanish and translated the Aztecs' words. Some called her La Malinche.

From the villagers he heard awesome stories of the power of Moctezuma and the Aztecs. With Doña Marina's help, he learned that these people—who were mostly farmers—hated the Aztecs. (You'll soon learn why.)

Stand with Cortés and his 400 soldiers. They are approaching the Aztec capital. They shiver as they march through a high mountain pass. There is snow underfoot, even though they are between two volcanoes. One of the volcanoes, Popocateptl (poh-puh-ka-TEP-tul), is spitting smoke and flame and bits of rock and ash. It roars a welcome—or is it telling the invaders to leave?

Cortés is not about to leave. As he nears the city he rubs his eyes. He can hardly believe what he sees. Tenochtitlán is more beautiful than any city on earth, he says. It is an island city, five miles square, surrounded by a glistening lake. Canals, filled with canoes, are used as streets. Three great causeways—earth bridges—lead in and out of the city. Bridges on the causeways can be raised to keep the capital safe from invaders. More than 350,000 people live in Tenochtitlán. They are artisans, warriors, priests, merchants, and government officials. Farming is done on the surrounding lands.

Cortés and his men are dazzled. The lake shines turquoise in the morning sun. Houses and public buildings are chalk-white or earth-red. Some are gilded, as if made of the gold the Spaniards covet.

There is more to see: gardens floating in the lakes, houses with patios and fountains, and, in one of Moctezuma's palaces, a private zoo. The market amazes the Europeans: 25,000 people are in the market

Cocoa Nuts

According to an ancient tale, Quetzalcoatl stole the cocoa tree from his brother and sister gods, gave it to the Toltecs, an Aztec people, and taught them to make chocolate. Wherever it came from, chocolate was prized. The Aztecs demanded cocoa from other peoples as a form of tribute. During the reign of Moctezuma, almost 50,000 pounds of cocoa beans were brought to Tenochtitlán every year. Sometimes the beans were used like money; sometimes they were offered to the gods.

The Aztecs also used cocoa beans to make a frothy drink. The beans were ground in a mortar with corn kernels. Vanilla, honey, chili peppers, and spices were added to the mixture; then it was whipped with water. Ordinary people could rarely afford to drink chocolate. It was a drink for the wealthy.

Indians who hated the Aztecs helped the Spanish build ships small enough to sail into Tenochtitlán. The ships were carried over the mountains to the lake. Their cannon fire was deadly.

square, buying and selling. A pyramid, the Templo Mayor, towers over the scene. Flowers are everywhere. So are birds. Thousands of herons, parrots, hawks, and egrets squawk and chatter from huge cages. No city they have seen has prepared the Spaniards for this place. Madrid, Spain's largest city, smells like a sewer. Tenochtitlán is not only splendid, it is clean.

The Aztec empire is glorious. It encourages art, music, poetry, and crafts. But it has a terrible flaw: the flaw is a religion that demands the sacrifice of thousands of people each year.

The Aztecs believe in other gods besides the peaceful Quetzalcoatl. They fear some of their gods and believe they demand what is most precious—life. So they kill people and give their hearts to the gods in religious ceremonies. They think the gods will bring earthquakes and other disasters if they aren't fed enough lives. It is the sons and daughters of their neighbors that the Aztecs sacrifice to the gods. You can now understand why those neighboring peoples want to help Cortés.

At first he doesn't need help. The Aztecs give him everything he wants. But Cortés has been sent to conquer, and he knows only one way to do that: with strength. Besides, there are just a few Spaniards in this empire of Indians; the Spaniards believe they must be ruthless to survive. Do you think they are right? What would you do?

Cortés captures Moctezuma and holds him hostage. Then, because Cortés intends to conquer this nation, he fights. He fights brilliantly. Of course, he has guns and the Indians don't. They have never even seen guns before. And there is something else: the Europeans have brought smallpox germs with them. When the Indians catch smallpox, they usually die. Those who are left are weak and sick. A terrible epidemic rages in Tenochtitlán, yet the Aztecs fight on. When Cortés tries to stop the killing, the Aztec leaders refuse to surrender. They are proud and vain. They are willing to die and see their city ruined rather than give in. And that is what happens. Soon there is almost nothing left. Tenochtitlán is burned down and pushed into the lake. Moctezuma is dead, and so are most of his followers. The Indians who are left are servants of the Spaniards or lead a primitive existence.

Artisans are craftspeople: potters, weavers, metalworkers, woodcarvers, basketmakers.

Little remains of the complex, brilliant Aztec culture except souvenirs.

Cortés builds a Catholic cathedral where an Aztec temple stood. He fills the lake with earth, proclaims Tenochtitlán a Spanish possession, and calls it Mexico City. He sends Mexican gold and silver back to Spain—boatloads and boatloads of it. From Mexico, which is now called New Spain—Nueva España—the conquistadors subdue South America and explore North America from California to Virginia.

Why did they destroy a great empire? Why did they steal a nation's riches? Were the Spanish evil and ruthless? Or were the times so different that it is difficult for us to imagine them?

Life in the 16th century was cruel; and punishment was often swift and horrible. That was true all over the world—in America, in Europe, in Asia, and in Africa. The piles of skulls in Tenochtitlán—left from the sacrifices—horrified the Europeans. They said that was the reason they had to destroy the Aztec empire. Was it a good reason—or just an excuse? In European cities criminals were hanged and left to rot in public view. That would have horrified the Aztecs. Suppose Cortés had come without his guns. Would he have lived? Do we have injustices and cruel practices in our country today? What are they? What can we do about them?

Reading history is not always easy. It is hard to make judgments about the past. But it is worth trying. It helps us make judgments about the world we live in.

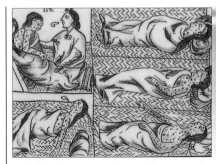

The Europeans have an immunity to smallpox. It can make them sick, it causes some of them to die, but most of them survive. The Aztecs, on the other hand, die in thousands.

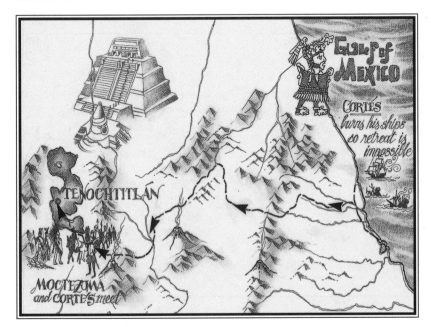

GULF OF MEXICO

CORTÉS burns his ships so retreat is impossible

TENOCHTITLÁN

MOCTEZUMA and CORTÉS meet

Mexico is an Indian word. Tenochtitlán was also known as Mexico, the City of Mexico, or Mexico City.

Tenochtitlán: "Place of the Prickly Pear Cactus Fruit"

These words are by a conquistador, Bernal Díaz del Castillo, who was with Cortés in Mexico. His book is called The Conquest of New Spain.

All about us we saw cities and villages built in the water, their great towers and buildings of masonry rising out of it. On dry land were other great towns, and with the straight, level causeway leading toward Mexico it seemed like the enchantments they tell of in legend. Some of our soldiers even asked if it were not all a dream.

At Iztapalapa, half the houses stood in the water, half on dry land. As we approached, splendid chiefs came to meet us, with a [magnificent] present of gold.... They lodged us in spacious palaces of beautiful stonework and fragrant woods. There were great rooms, and courts wonderful to behold, all covered with awnings of cotton cloth. We wandered through gardens where I never tired of looking at the different trees, each with its own scent. There were paths full of flowers and there were orchards and ponds. A lake of clearest water was joined to the grand lake of Mexico by a channel capable of admitting the largest canoes. All was ornamented, painted, and admirably plastered, and delightful with singing birds. When I beheld the scenes around me I thought within myself, this was the garden of the world. And of all the wonders I beheld that day, nothing now remains. All is overthrown and lost.

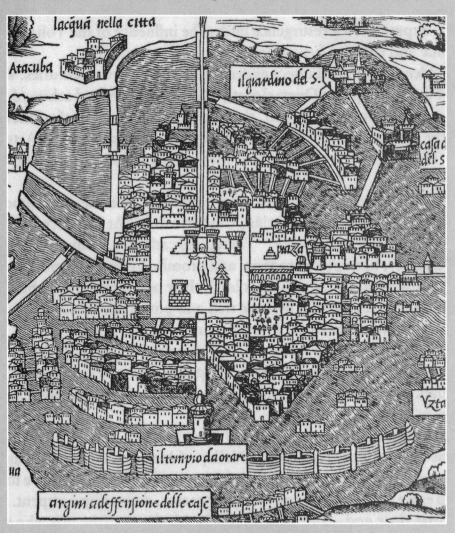

Tenochtitlán was built on an island in Lake Texcoco, one of five lakes in the Valley of Mexico.

23 Ponce de León, Pizarro, and Spanish Colonies

The Incas valued the alpaca's fur too much to eat its meat. This model of an alpaca is silver.

Juan Ponce de León (hwahn PON-say day lay–OWN) heard the stories of Cortés and his great success. He believed he, too, could find kingdoms of gold, and he thought he knew just where to look.

Ponce de León had come to the New World with Columbus, but he was different from Columbus: he was tougher. Some say he was bloodthirsty and cruel, but he was also brave and generous. He gave most of his money to charity.

Ponce de León was related to Spanish kings. He was a page in the royal court of Spain when he set out to find adventure in the New World. On the island of Puerto Rico he found more than adventure: he found gold. He conquered Puerto Rico, became its governor, and made a fortune in gold, slaves, and land. But Ponce de León wasn't finished exploring. He heard tales from the Indians of a magical spring that cured illnesses and made old men and women young again. He set out to find that Fountain of Youth.

He failed to find it. Instead, in 1513, Ponce de León discovered a new land. The new land, which was North America, was filled with beautiful flowers. Ponce de León called it La Florida. In Spanish, *florida* means "flowery."

But Ponce de León still wasn't satisfied. He wanted to surpass Cortés. He wanted to find something even greater than Moctezuma's kingdom of gold, and he was convinced he would find it in Florida. In 1521 the king commissioned him to conquer and colonize "the island of Florida." (No one knew it was more than an island.) Ponce de León set out from Puerto Rico with two ships, 200 men, and 50 horses. Unfortunately for him, all he found in Florida were Indians who shot poisoned arrows. One arrow entered

El Dorado—in Spanish it means "the golden one"—was a long-lived legend among the Europeans who first reached the New World. The Spaniards believed that somewhere in the Americas was a place where gold was as common as sand. For years explorers sought it fervently. Sometimes *El Dorado* meant the king of this mythical land; sometimes it was the place itself. Many died looking for it.

Ponce de León, "brave lion."

Even on foot, messengers traveled fast on the excellent Inca roads. This one blows a conch shell to proclaim his arrival.

Land of Gold

The words of the 16th-century German artist Albrecht Dürer, on seeing gold objects from the Incas:

Then I saw the things which were brought to the king out of the new Land of Gold...all sorts of marvelous objects for human use which are more beautiful to behold than things spoken of in fairy tales....and I marveled over the subtle genius of those men in strange countries.

Ponce de León's thigh, and the poison began to work. When that happened his men fled from Florida back to Cuba, where the tough explorer died. He was buried under a stone that says, "Here lie the bones of the brave lion." (In Spanish, *león* means "lion.")

However, Ponce de León was right. There was another kingdom of gold—it just didn't happen to be in Florida. Francisco Pizarro (remember that name?) headed down the west coast of South America and found golden treasures beyond anything anyone had ever imagined. He found them in Peru.

Pizarro's capture of Peru, in 1532, was perhaps the most daring and terrible of all the Spanish conquests. With just 180 men, 67 horses, and three big, noisy guns, Pizarro defeated the powerful Inca empire.

The Incas fought to save their city from Pizarro (above) and his men. But they were beaten by guns and greed.

98

When Pizarro arrived in the Inca capital, Cuzco, he captured the ruler, who was known as the Grand Inca. Pizarro promised to release the Inca, whose real name was Atahualpa (at-tah-WAL-pah), if his followers would fill a huge room with gold. They did, but Pizarro killed Atahualpa anyway.

As you can see, Pizarro and his men were not exactly honorable. In fact, they were deceitful and treacherous. Soon they were fighting among themselves for gold and power. They ended up killing each other. Pizarro was killed, too. Some say Atahualpa's ghost got revenge on Pizarro. Things got so bloody that finally the King of Spain took over. He didn't mind at all. Spain was going to grow rich on the gold and silver from the mines in Peru.

To ransom their imprisoned ruler, Atahualpa, the Incas brought enough gold objects to fill a room that may have been 22 feet long by 17 feet wide.

Sometimes historians say they want to cry when they think of Pizarro's conquest. Remember that room the Incas filled with gold? Well, it wasn't just gold. It included beautiful jewelry and carvings of animals and birds and decorated plates and spoons—the artwork of a civilization. The silver alpaca pictured at the beginning of this chapter was saved. But much more was not. Pizarro melted down all the gold into bars. It was lost to history forever.

The Spaniards did that kind of thing many times over. Their religion told them the Indian civilization was pagan and therefore false, and that its symbols should be destroyed. Because they believed their religion was the only true religion, they thought they were doing the right thing when they forced it on others.

Spain and the other European nations had guns, powerful crossbows, ships that could sail into the wind, and printing presses that made the exchange of ideas easy. Sometimes they acted as if that strength gave them the right to bully other peoples. Some Europeans said, "might makes right." Many Spaniards believed that their nation was best because it was strongest.

A few people questioned those ideas, but most did not. When leaders say something is all right, most people agree, without thinking for themselves.

A ***pagan*** (PAY-gun), in Pizarro's time and until very recently, was anyone who was not a Christian, Muslim, or Jew.

24 Gloom, Doom, and a Bit of Cheer

Europeans had figured out how to sail to America, but not how to cure bad diseases.

Great Plague

The Black Death raged through Europe and Asia in the 14th century and killed more people than any war. It began in China and spread west. One estimate says that at least 40 percent of the population of Asia, Europe, and North Africa died in 20 years. About 90 percent of those who got the plague died. The same kind of thing happened when smallpox germs came to America. Today, with antibiotics, the death rate from these diseases is about 5 percent.

Historians guess there may have been at least 20 million Indians in Mexico when the first Spaniards arrived. No one knows just how many died after the Europeans came, but almost everyone agrees it was more than three-quarters of the Indian population. Imagine: if you were a Native American living then, three out of every four people you knew would be dead.

It was an accident that did most of the killing—the accident of disease. No one intended it. Millions of Indians were killed by the germs that came with the Europeans and Africans. Some say only two million Mexican Indians survived. European diseases had a similar effect on Native Americans everywhere. Few people understood the importance of cleanliness or how to combat infections. People were used to epidemics. They just shrugged their shoulders and called them "God's will." In the 14th century Europe suffered a terrible plague called the Black Death. People looked on helplessly as one out of every three Europeans died.

So the European adventure in the New World began with this terrible accident of disease and death. But there were other things, that weren't accidents, that also made life difficult for the native peoples. The Europeans came to America to expand their world and to enrich themselves. They came as conquerors and colonists.

A colony is a region controlled by a foreign country. The conquerors saw themselves as parents and their colonies as children. Spain became a "mother country." Mexico was its colony. (The Mexicans didn't think they were children, but no one asked how they felt.)

Mexico was ruled by Spain in a way that was good for the parent, but not always good for the child.

It wasn't all bad, though. Spaniards brought their religious faith and their architecture to Mexico and South America. They brought their language, their arts, and their elegant manners. They brought learning: the first printing press arrived in Mexico City in 1539, and a university in 1551. They encouraged truthtelling: they let their historians write the good and the bad about what was happening in America. They built magnificent churches and palaces. They ended the terrible blood sacrifices that had been part of the Indian religion in Middle America.

They tried to make America Spanish. But the nations they created were neither Spanish nor Indian. They were a hybrid; that means "a mixture." Spaniards married Indians, and their children were called *mestizos*. Spaniards married black people, and their children were *mulattoes*. Soon all these people were living together in Mexico, Peru, and other Spanish colonies.

The Spanish colonies were not friendly with one another. Back in the 15th and 16th centuries people in different parts of Spain didn't get along very well. In America their differences grew wider. The Spaniards never built good roads from one colony to another. (That would have been difficult to do with the many mountains and jungles about.) That is one reason there never was a United States of South America. Today South America is filled with separate Spanish-speaking countries.

But the people in the largest country in South America do not speak Spanish. What country is that?

It's Brazil. Brazil became a Portuguese colony, and Portuguese is the Brazilian language. In three small countries on the northern coast of South America, three other languages are spoken. See if you can discover the names of those countries and their languages.

It was Spain, however, that dominated South America. The Spanish might have made all of North America a colony if they had found gold in the north. They didn't, although they tried hard enough.

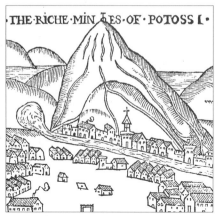

THE·RICHE·MIN ES·OF·POTOSS [·

Europeans poured into the Americas in search of wealth. Sometimes, as in Peru's silver mines at Potosí, they found it. Some found riches in tobacco or sugarcane instead. But many more just died still searching.

The life of many Incas born after the Spanish invaded—like the young noblewoman on the left—was very different from their grandparents' lives. It was much poorer and harder.

25 North of New Spain

Water moccasins like this were only one of the problems in Florida's swamps.

Everyone knew there was gold in the North—there had to be. The sages in Europe said so. Things of the world come in groups of three, they said. Gold had been found in Mexico and in Peru; it would also be found in the north. There would be no balance without it, so they said.

And the seven cities of Cíbola (SEE-bow-la)? They were in North America, said the wise men of Spain. Europeans had been searching for the seven cities for a long, long time. Tales were told of seven priests who had wandered off from Europe and founded seven fabulously rich cities.

Explorers had looked for the cities in Africa. Then they thought they might be in the Orient. But now everyone was convinced they were in North America, especially when they heard the Indians talk of seven cities. (The seven Indian cities turned out to be pueblos, built of adobe clay, not gold. Some Spaniards saw beauty in the pueblos; for most they were a terrible disappointment.)

Indians panning for gold. They scooped up a panful of water and sand. When the water settled, the heavy particles of gold separated and sank to the bottom of the pan.

Oh, how they tried to find the seven cities of the legends!

The governor of New Spain (Mexico) sent a team of men to find the cities. He thought he had the perfect explorers when he picked Estebán (ess-tay-VAN) and Fray (FRY) Marcos de Niza. But some thought they made a strange pair. Fray Marcos was pious and serious. He didn't know what to make of Estebán, an African from Morocco who had success as a healer and liked to dance and sing. The Indians thought him a great doctor.

Estebán had come to America as a slave in 1528, with 400 Spaniards who landed in Florida in search of gold. They were led by a one-eyed, red-bearded conquistador named Panfilo de Narvaez (pan-FEEL-o day nar-VY-yez). Narvaez was rich, disorganized, and horribly cruel. He had lost his eye fighting Cortés. Narvaez marched his men up the west coast of Florida. It was alligator country, and the Spaniards must have heard those beasts bellowing. But Alvar Nuñez Cabeza de Vaca (NOON-yez ka-VAY-sah day VA-kah), who survived and wrote of the expedition, never mentioned alligators, or the poisonous snakes that slither through the region. He did note the ducks, geese, herons, deer, rabbit, bears, "lions" (which may have been cougars), and an "animal with a pocket on its belly, in which it carries its young." The kangaroo? No, it was the opossum, unknown in Europe.

Indians "playing flutes of reed" serenaded the explorers. It was a traditional form of greeting. But Narvaez was no music lover. He had a few hundred Indians killed—for no special reason. Then he forced some Indians to take him into the interior of the land. There the Spaniards captured an Apalachee Indian village and discovered corn, cornfields, and "fine pastures for herds."

By this time Narvaez and his men had had enough of Florida. In addition to Indians they had been fighting huge mosquitoes. And there didn't seem to be gold or golden cities anywhere. They set off for the coast in search of their ships.

Meanwhile, some natives were watching the invaders. Cabeza de Vaca said they were seven feet tall, with bows "as thick as the arm" and arrows capped with snakes' teeth. The Indians were waiting for the right moment. It came when the conquistadors were in deep water crossing a lake. Poisoned arrows rained down upon them. The sharp arrows cut right through their suits of woven chain mail.

The surviving Spaniards couldn't find their ships. They were starving and desperate. So they built five boats. Then they ate their horses, made the horsehide into water bottles and sails, and pushed off. Since they didn't know much about boat-building, most of their boats sank.

Of the 400 men who had come to Florida, Estebán was one of only

A German soldier, Ulrich Schmidel von Straubing, went to America with Cabeza de Vaca and wrote a book about their adventures.

Fray means "friar" in Spanish. A friar is a member of a Catholic order of holy men—like Friar Tuck, Robin Hood's companion.

103

A MAP of the WEST-INDIES &c. MEXICO or NEW SPAIN. Also ÿ Trade Winds, and ÿ severall Tracts made by ÿ Galeons and Flota from Place to Place. By H. Moll Geographer

Two hundred years after the journeys of Cabeza de Vaca and Estebán, this map was drawn to show the routes of the Spanish treasure ships in the West Indies. This land was the heart of New Spain.

Ceuola, an Indian word, sounded like *Cíbola* to Fray Marcos.

four who lived. (Cabeza de Vaca was another.) Estebán made it to land (probably at Galveston Island, Texas), where he was captured by Indians and learned their language. When he finally got to Mexico, eight years later, he had walked across a lot of the land that was to become the United States. That's why the Spanish governor of Mexico thought he was a perfect choice to send out on another exploring mission—after gold again, naturally. Fray Marcos was to go with him.

Fray Marcos had a hard time marching with Estebán because Estebán had so many women friends. That was upsetting to the Catholic priest. So Fray Marcos suggested that Estebán go ahead as an advance scout. "Send back a small cross if you hear of gold or of a great city," the friar said. "Send a medium-size cross if there is much gold."

Estebán was like the Pied Piper: he attracted followers. He was soon leading a joyful procession of 300 Indian men and women. He wore turquoise jewelry around his neck and feathers on his ankles and wrists. He sent Fray Marcos a huge cross "as high as a man." And he sent a messenger who told of a mighty city the Indians called Ceuola. Could it be Cíbola?

But another messenger soon came running to Fray Marcos in tears. Estebán was dead. No one could believe it! Most Indians had loved Estebán. He thought everyone would love him. But when he sent a gourd with a red feather on it as a greeting to the Zuñi, he was killed. The Zuñi had heard stories of the Spaniards' cruelty. Besides, the red-feathered gourd was their symbol of war.

Fray Marcos went back to Mexico City to report. He told of golden cities and said Ceuola was "the greatest thing in the world...larger than the city of Mexico." He was sure he had found the seven cities of the legends.

You can imagine how excited people were in Mexico City. Everyone wanted to leave at once. They wanted to find Fray Marcos's Cíbola, the City of Gold.

To kill an alligator, Indians first shoved a pole down its throat. Then, wrote the artist LeMoyne, "the beast is turned on its back and killed by beating it with clubs."

Looking for Cíbola

This is Fray Marcos's report about Cíbola, as quoted by Richard Hakluyt in his Collection of the Early Voyages, Travels, and Discoveries of the English Nation. *Hakluyt was an English geographer who wrote about America in 1582. The name "Stephan" is English for* Estebán.

So the said Stephan departed from me on Passion Sunday after dinner: and within four days after, the messengers of Stephan returned unto me with a great cross as high as a man, and they brought me word from Stephan, that I should forthwith come away after him, for he had found people which gave him information of a very mighty Province, and that he had sent me one of the said Indians. This Indian told me that it was thirty days' journey from the town where Stephan was, unto the first city of the said province, which is called Ceuola.

26 Being a Conquistador with Coronado

This is the oldest European drawing that there is of an ear of corn.

Estebán and Fray Marcos had failed to find gold. But they were a strange pair. So when people talked of Cíbola, they said the wrong men had been sent to search for those fabled cities. What was needed was a great leader.

Everyone knew that Cíbola existed. No one questioned that. The stories of its riches grew and grew—and grew. It was as rich as the Inca empire. No, it was much richer, people said.

"Haven't you heard? People in Cíbola use gold pots to cook their dinners." That was what was said and believed.

Francisco Vasquez de Coronado (VAS-kes day kor-oh-NAH-do) would find Cíbola; everyone agreed about that. He was a fine soldier and an able leader of men. So when it was decided that Coronado would lead an expedition, men and women were eager to go with him. It was to be an expensive venture, and well planned, which was the way Coronado did things.

The year is 1540. Coronado has gathered his followers together near Mexico's Pacific coast. He is not yet 30, blue-eyed, with blond hair, a mustache, and a beard. He is preparing to set out on one of history's great adventures.

We are with him. If we choose to be Spaniards, we will wear plumed helmets and shining armor. That doesn't make much sense on hot days, but it does look good. Some families have come along, so some of us are women and some are children. Most of the 300 of us who are Spaniards are on horseback. Can you feel the excitement? We expect to find riches greater than those found by Cortés, or even Pizarro. We will be exploring lands new to Europe. We will be making history.

A few of those among us who are Indian or black may ride horses

Maize Miracle

Maize is Indian corn. It is good to remember that two American products have probably had more influence on world history than anything else (even the atom bomb). Those products are corn and potatoes. After Europeans reached America and brought back the plants they found, these two foods became staple (basic) crops for many Old World countries. Can you figure out why? (Hint: wheat needs very good soil and rice needs a lot of rain.)

Boats couldn't always get close enough to shore so people—or horses—could walk onto them over gangplanks. These horses were hoisted on board with ropes.

back, but most of the non-Spaniards have to walk. There are some black people and more than 1,000 Mexican Indians on this expedition. Some explorers treat Indians like animals, making them carry heavy loads. Coronado is not like that: he has mules and horses to carry equipment. Animals also drag heavy cannons. Coronado wants to be prepared for any emergency.

Of course we will get hungry. So herds of cattle and sheep—thousands of them—come along. They will make juicy barbecues.

The people on foot and the animals travel slowly. Coronado and his conquistadors become restless and ride on ahead. Fray Marcos is with them as a guide.

Look at a map. Do you see the southeastern corner of Arizona, right next to New Mexico? Do you see the city of Douglas? That is just about where Coronado goes. Then he heads due north, for Cíbola. Take a finger and slide it up the border between the two states. When you hit Zuñi, in New Mexico, you may stop. Coronado stopped near there, too, at the Indian village of Hawikuh—the very village where Estebán was killed. Can you believe what you see! This is supposed to be one of the seven cities of Cíbola! But Hawikuh is a pueblo, not the city of legend, not the golden Cíbola. Coronado calls it a "little, crowded village...all crumpled up together."

"Such were the curses that some hurled at Fray Marcos," a soldier says, "that I pray God may protect him from them."

By this time a year has passed. Food has run out. Everyone is hungry. One of the Spaniards writes, "We found something we prized more than gold or silver—namely, plentiful maize and beans, turkeys, and salt." To get that food, the Spaniards kick the Indians out of their pueblo. The battle doesn't take long. The Indians fight with rocks and arrows; the Spaniards have guns and crossbows.

Coronado hasn't given up. He is still sure he will find Cíbola. But it isn't easy. There are no maps to tell him where he is. Everywhere he

The word **barbecue** means an outdoor roast. It comes from a Taino word—*barbacoa*—that means "an arrangement of sticks." The sticks were used to hold meat above an open fire.

Pedro de Castañeda, one of Coronado's soldiers, wrote of the Zuñi Indians of Hawikuh: "They are ruled by a council of old men. They have priests...whom they call *papas* (elder brothers)... there is no drunkenness among them...nor sacrifices, neither do they eat human flesh or steal, but are usually at work."

"It was impossible to descend....What seemed very easy from above was not so, but indeed very hard and difficult."

looks, the grass is as high as the horses' bellies. Coronado says it is like being at sea. He uses a compass to guide him.

One exploring party heads west, to Arizona, with García Lopez de Cardenas (gar-SEE-yah LOW-pez day kar-DAY-nass) in command. They come upon a canyon so vast that some have called it a mountain in reverse. At its deepest it is a mile down; at its widest it is ten miles from rim to rim. The canyon—with stone towers that look like castles or ships or enchanted creatures—has a tumbling river at its base. This is the Grand Canyon. The conquistadors try to reach its bottom. They get partway down and give up. It is too deep and too dangerous.

Other scouting parties find the land teeming with deer, elk, bear, beaver, snakes, and birds. Ponderosa pines forest the mountains. Purple iris, white mariposa lilies, and yellow columbine poke through the grass.

The sun reflecting on the icy mountain peaks is so bright that it hurts the explorers' eyes. In Europe the landscape is small-scale. There, trees and hills and houses block the view. Here, eye distances are vast. Broad plains stretch on and on. Mountains rim the plains; their rocky layers tell a story of volcanoes and mountain uplift. The whole scene changes color with the time of day: from purple to orange to ocher to tan. In a landscape such as this your eyes can play tricks on you. Distances seem shorter than they really are. Men and women feel small as ants. The conquistadors don't like it at all.

And there are dangers. A horse stumbles in quicksand and is soon submerged. The riders complain of the high grass. Too bad—in a few hundred years the grass will all be gone, trampled by horses, sheep, and cattle or suffocated by European weeds. But that does not concern these people; their mission is to find treasure. Most have invested everything they own in this expedition. They have not come to look at scenery. "Where, oh where is Cíbola?"

Do you have an idea now of what it was like to be a conquistador? You do? Well, whatever you think it was like, it was much worse. It was too hot or too cold or too exhausting or too discouraging. There was hunger and sickness and fear of strange animals. Anyone who wandered from the main expedition was likely to get ambushed by Indians. Look at the map again. Those are great distances. It is a long way to go from Mexico to New Mexico on horseback—and even longer on foot.

Some of the North American Indians figure out a way to get rid of the Spaniards. They tell them of golden cities that are always a few days' march away. One Indian whom Coronado captures is nicknamed

"the Turk." Since many Spaniards still believe that China and Turkey are somehow connected to this continent, they may really think he is a Turk.

The Turk can't speak Spanish; he communicates through sign language. But the Spaniards are sure he is telling them of a land ruled by a king who dresses in silk. The Grand Khan of Cathay dresses in silk. Things are beginning to sound exciting.

The Turk says the word "Quivira" (kee-VEER-ah), and they know it is a city. They are certain he says it is built of gold. They go to find it.

The Turk heads northeast, and they go with him, a long, long way, to the land that is now the state of Kansas. Finally, they reach Quivira. It is a city of mud huts. Can you imagine how they feel? Look at that map again. Think how tired you are if you are a Spanish conquistador.

The Turk has been lying, or so the conquistadors believe. They strangle him. (Those conquistadors didn't fool around.) Now Coronado and his party head home, discouraged and disappointed. They have been gone for two years and have traveled more than 7,000 miles. They have marched over land that holds gold and silver and copper—but it will take future generations to discover that.

You will read about de Soto in the next chapter. (You will read about Cabrillo, too.) De Soto met Indians who said they had seen other white men. But de Soto never found out that he and Coronado had come within a few hundred miles of each other.

It's 50 years since Columbus went looking for the Indies, and yet the Spaniards are still using that out-of-date information about the Grand Khan from Marco Polo. It took a long time for new ideas to get around in those days.

27 Conquistadors: California to Florida

The griffin was a mythical animal, part eagle, part lion. Explorers expected to find griffins in the New World.

The race was on! Whoever found Cíbola would become rich and famous. If you lived in the 16th century, would you have set out to find Cíbola? Many Spaniards did just that. They were willing to risk everything in the hope of finding the treasured cities.

A decade is ten years. During the decade of the 1540s—from 1540 to 1549—while Coronado was searching in Kansas, some Spaniards went looking in California for the seven golden cities. Others looked in places from Florida to Texas.

In the summer of 1540, Hernando de Alarcón (air-NON-do day al-ar-KON) sailed up the Gulf of California, between Mexico and Baja (BY-yah—it means "lower") California. He thought Baja California was an island and that he would meet up with Coronado on it or on the ocean passageway he was sure he'd find north of Mexico. But when Alarcón landed at the mouth of a river, he realized that California was no island. What he didn't know was that the river was the same one that flows through the Grand Canyon: the Colorado River. Alarcón went up the Colorado River, traded with Indians, galloped around, and searched for Cíbola. When he didn't find the golden cities—or Coronado—he turned back to Mexico. No one is quite certain, but Alarcón may have been the first European to reach what is now the state of California.

We know for sure that another conquistador got there. He was Juan Rodriguez, although his men called him *cabrillo*, which means "little goat" in Spanish. Soon everyone was calling him Cabrillo: Juan Rodriguez Cabrillo (hwahn rod-REE-gez cab-REE-yo).

Cabrillo sailed up the Pacific coast from western Mexico and, after three months, came to what we now call San Diego Bay. It was 1542, exactly 50 years after Columbus landed in the New World. The Spaniards had discovered California.

The name California came from a book that told about a wild land, called California, at "the right hand of the Indies." This land was full of gold and treasure but inhabited by griffins: fierce flying creatures that ate people. We think of griffins as mythological (or imaginary) beasts. But many Spaniards, at that time, believed they were real. It took courage to be a conquistador.

Cabrillo and his crew sailed up the California coast. The pilot of his ship described what they saw: "There are mountains which seem to reach the heavens, and the sea beats on them; sailing along close to land, it appears as though they would fall on the ships."

The mountains didn't fall, but Cabrillo did. It's too bad he wasn't a little goat. Chased by Indians, he slipped on some rocks, hit his head, and died. Being an explorer was a risky business.

Back in Spain, Hernando de Soto (air-NON-do day SO-toe) heard Florida described as a "land of gold." De Soto had been in Peru with Pizarro and had become immensely rich. When he decided to explore La Florida, there was much excitement in Europe. Men sold all their possessions to buy a place on his expedition. An engineer came from Greece, a longbowman from England, and four fortune seekers from Africa. From Spain came knights, priests, artisans, and adventurers. More than 500 people signed up; two were women.

De Soto was an admirer of Atahualpa and one of his advisers, too. He was horrified when Pizarro killed the Inca ruler.

All of them expected to find wealth and fame. Besides gold—and Cíbola—they were also looking for a waterway that would take them on to China and Japan. No one had any idea of the size of the land they called Florida. They thought it was an island, like Cuba, only bigger. It seemed logical that Japan was the next island they would find. No one guessed that the lands Cabrillo and Coronado were exploring were part of the same continent as Florida.

It is 1539. Let's watch as de Soto, his men, and their horses clatter off the boats in Florida. They are in a holiday mood. De Soto is a brilliant leader; he has trained them well and kept them enthusiastic. Their armor shines, the horses are frisky, flags fly overhead, and all are anxious to use their weapons and show how brave they are. It isn't long before they have a chance. They learn of an Indian village and, according to an expedition member, "gallop their horses...lancing every Indian encountered on both sides of the road."

Since they have guns and horses, and the Native Americans don't, they capture an Apalachee village. (As you may remember, the unlucky Apalachees have already been attacked by Narvaez and his men.) The Spaniards find corn, beans, and pumpkins, and have a feast. But the Indians, who are "very tall, very valiant, and full of spirit," don't give up easily. They set fire to the village and ambush Spaniards who wander away from the camp. It is a tactic that will be used against Europeans as long as they fight Native Americans.

The Europeans do best when they form a broad line of men on horseback and charge their enemies. That is how they fight on European battlefields: armor gleams, banners fly, horses snort, and trumpets cry. It is a glorious sight, meant to frighten enemies. But it doesn't work in the thick woods and swamps of North America. There are no open battlefields. The Indians use guerrilla (say it like *gorilla*) methods: arrows come from unseen men hiding behind trees. De Soto and his soldiers will try to adapt.

Right away they have a piece of good luck. They are about to slaughter a group of unarmed natives when one shouts to them in Spanish. He is a survivor of the Narvaez expedition who has been living as an Indian for 11 years. Because he now speaks Indian languages, he will interpret for de Soto. The Spaniards will be able to talk to the Native Americans.

But de Soto has come to conquer, not talk. He is about as kind as a ravenous panther; he encourages the Spaniards to torture, burn, and kill the Indians they capture. He has brought a pack of snarling attack dogs with him and he throws captives to the dogs. Word of his brutality spreads from tribe to tribe. Some Indians flee as de Soto and his men march through the country, but others give him the food and slaves he demands. They have little choice: if they don't give him what he wants, de Soto takes hostages. Usually he captures Indian chiefs.

De Soto has made careful plans. He has brought 13 hogs, and he breeds them. After a year there are 500 swine. His men will not go hungry.

Picture this procession through the woods. First come proud conquistadors on horseback holding long lances; sturdy foot soldiers march behind them. Then come African slaves; then, Indian guides. Next come chained Indian slaves hauling supplies; then, squealing pigs and their herders, and, finally, more conquistadors.

De Soto and his followers slog through swamps, cross rivers, climb mountains, and push through jungles and forests. They build rafts of wood and bridges of rope.

Marching through the South in springtime, they find "countless roses growing wild" and strawberries, "very savory, palatable, and fragrant." The soldiers eat wild spinach—although they prefer meat—

A **lance** is a weapon like a spear, with a long wooden shaft and a metal head.

Savory means "tasty"; so does **palatable**—or eatable, at least.

Marche du Calumet de Paix.

The leader of these Natchez has a special ceremonial pipe called a *calumet* (KAL-yoo-met). It is a peace pipe. Those are Europeans in the shelter. They will smoke the pipe of peace with the Indians.

and they learn to enjoy corn cakes as the Indians make them, with "oil from walnuts and acorns." They collect pearls—200 pounds of them—mostly by robbing tombs.

In Alabama they are greeted by an Indian chief carried on a cushioned, ceremonial bed. The chief wears a crown of colorful feathers and a gorgeous fur shawl. He is surrounded by attendants who are "playing on flutes and singing."

The Spaniards are impressed. These are not savages. One of the explorers writes about what they find: "The country is thickly settled in numerous and large towns, with fields extending from one to another....In the woods were many plums... and wild grapes...."

They are treated well; perhaps their guns give the natives no choice. A few men—African slaves and unhappy Spaniards—slip away and join Indian tribes. But the others act like what they are: conquerors. Then there is a catastrophe. Indians burn their camp. Most of the Spaniards' supplies—clothes, armor, guns, horses, pigs—go up in flames. The smell of fried bacon fills the air. But de Soto is at his best when faced with disaster. The conquistadors make clothes of dried grass and animal skin (and endure the laughter of their comrades). They make a blacksmith's forge. Lances, swords, and guns are repaired. The adventurers march on.

A *catastrophe* (ka-TASS-troh-fee) is a disaster.

113

Spanish conquistadors such as de Soto were called *adelantados*. It means "advancers." The Spaniards didn't advance the Indians' life much. Here de Soto and his men surround an Alabama Indian village.

Finally, de Soto and his men (both women have died) reach America's greatest river, the Mississippi. It's an important discovery, but they don't care about rivers. It is gold they seek. After almost two years of exploring, they have found nothing that they value. De Soto will not stop. The river is almost two miles wide and full of treacherous logs and huge fish. (Some Mississippi catfish weigh more than 100 pounds.) Indians, sitting in canoes, watch them. De Soto builds four boats and ferries men, horses, and hogs across the broad Mississippi.

Now on the west side of the river, they head on: through marshes, bogs, canebrakes, and bayous (swamps) where resting alligators look "like stumps of trees." In Texas (or what will be Texas), Indians tell them of other men who speak as they do. The Native Americans are describing Coronado and his men, although de Soto does not know that. (At this very time, in 1542, Coronado is on his way back to Mexico.) What if the two armies had met! They would have understood that New Mexico and Florida were part of the same land.

De Soto pushes on. He seems to think he is superhuman. But he is not.

He has been able to bluff and bully his way across half a continent. But when he catches a fever, it is all over. Hernando de Soto—the brave, mean, tough, brilliant conquistador—is dead. Now, his men are terrified. The Indians feared the cruel Spaniard and believed he had magic powers. Will they attack if they know he is dead? De Soto's men don't want to find out. At night they bury him in the middle of the river. Then they flee. They are happy to escape from the wilds of America. Of those who made this trip, 311 return to safety. They have been gone for four years. When they tell their story, most Spaniards decide that North America is not a very promising place.

Cane is tall, woody, spearlike grass. A *canebrake* is a thick growth of cane. It is hard to walk through a canebrake: the cane can cut your arms, feet, and legs.

114

28 A Place Called Santa Fe

Even the meanest conquistadors believed it was a duty to convert Indians to Christianity.

After the decade of the 1540s the Spaniards forgot Cíbola. Well, maybe they didn't forget it—they just gave up looking for it. Fifty years passed. Then, in April of 1598, they tried again.

Watch as eight men and their horses stagger to the Rio Grande (the great river that separates Mexico from Texas). They have been five days without food or water. Two of the horses drink so much, so fast, that their bellies burst and they die. The men get drunk on water.

They are a scouting party sent ahead by Juan de Oñate (hwahn day oh-NYAH-tay). Oñate's wife is the granddaughter of the great Cortés and the great-granddaughter of none other than the Aztec Emperor Moctezuma. Oñate has become rich by mining silver. If anyone can find Cíbola, it should be Oñate and his wife.

But since Cíbola doesn't exist, of course Oñate can't find it. He doesn't find gold either, although he will travel farther than Coronado did searching for it. What he does do is start a Spanish colony in New Mexico, a place where Spanish men and women make homes. "I take possession, once, twice, and thrice...of the kingdom and province of New Mexico," says Oñate. With him are 400 men, women, and children and 7,000 animals. Goods are carried in 80 wooden wagons with iron-rimmed wheels. The wagons are called *carros* and are heavy and expensive. Each is pulled by eight mules (with another eight tagging along in reserve). Oñate has paid for all this himself. It takes about a year and a half to make the slow trip from Mexico to New Mexico. Supply wagons will not come often.

Oñate is a dreadful man. He seizes food from the Acoma Indians. When they protest, he slaughters or enslaves them. That's just the

Like Oñate's wife, many Mexicans are part Spanish and part Indian.

Santa Fe is Spanish for "holy faith."
A **plateau** is a flat area, often in a high place or on top of mountains.

A **missionary** is a member of a religious group who goes out into the world to try to persuade others to **convert**—that is, to change their religion.

beginning. Things get so bad that most of the Spaniards are ashamed and horrified. But they are like sailors on a ship: Oñate is their captain, and they are not free to leave. Still, in 1601, when Oñate is off hunting treasure, some of them flee. Word of his behavior reaches the government in Mexico City. Finally, the Spanish court recalls him and takes away his title.

The New Mexicans start a new settlement on a cool plateau; they name it Santa Fe. It is 1610, and this is the first permanent European colony in the North American West.

The gold-seekers return to Mexico; their pockets are empty. Priests and settlers take their place. By 1630 the Catholic missionaries (they are Franciscan priests) say they have converted 60,000 Indians to Christianity. But many of these new Christians continue their traditional Indian religious practices along with the Catholic rituals. In about 1700 a priest named Francisco Eusebio Kino goes to Arizona. By this time the Indian population of the Southwest has been devastated. Diseases, brought from Europe and Africa, have almost wiped out the peoples. When Estebán and Fray Marcos searched for Cíbola, there were more than 100 pueblos in New Mexico. In 1700 there are only 18.

Spanish mission churches like this one in Taos, New Mexico, are some of the oldest European buildings in North America. You can visit them in Santa Fe, too, and in parts of Texas, California, and Arizona.

29 Las Casas Cares

Many Europeans believed it was right that some people should be slaves and others masters.

No doubt about it: Pizarro, de Soto, and Oñate were cruel. Yes, they were tough, energetic, and very brave; but they were cruel—and greedy, too. Now, if you are thinking that all the Spaniards were greedy and cruel, you're wrong. Many were kind, caring people.

The Spanish sovereigns cared. They worried about their subjects. (No, not things they studied in books. Subjects can be the people a ruler rules.) King Ferdinand and Queen Isabella believed the Native Americans were now their subjects. Queen Isabella made a law prohibiting the Indians from taking too many baths. She thought bathing was unhealthy.

In 1542 King Charles I—who was the grandson of Ferdinand and Isabella—ruled that Indians could not be made slaves. The king made his ruling because of a man named Bartolomé (bar-toh-loh-MAY) de Las Casas, who is someone to remember. Las Casas spoke out for human dignity and liberty as few people do. But that was after he turned 40. Before that he seemed to be like most other people.

When he was a boy his father and his uncle went on a great adventure. They sailed to the New World with Columbus. When Bartolomé's father returned, he brought his son a gift: it was an Indian to serve as the young man's slave.

What did Bartolomé think about that? How did he treat the Indian? What did he learn of the New World? We don't know, but we do know that in 1502 (which was how long after Columbus's first voyage?) Bartolomé de Las Casas sailed for the Indies himself. Soon after he arrived, he received a royal grant of New World land. The Indians who lived on the land were part of the grant. It was expected.

How could the king and queen of Spain give him land and people

A **sovereign** is a royal ruler, a king or queen or emperor.

Holy Roman Emperor

Charles I, or Carlos (the Spanish form of Charles), had several titles. He was also king of Germany and Emperor Charles V of the Holy Roman Empire. He ruled one of the largest territories in history: it stretched from Germany to Africa to the New World. The story of politics in western Europe in Charles's time is the story of his fights with Francis I of France, Henry VIII of England, Martin Luther, and the pope. It was a time of almost constant war.

117

Cierto no fue Díos servido de tan execrable injusticia is Spanish. It says, "Certainly God was not served by such awful injustice." Las Casas wrote this in *Historica*, his book about the New World.

in the New World? Well, they seemed to think their power gave them ownership rights. The Spaniards were conquerors (as were the English, Dutch, French, Portuguese, and Swedes later). They even used that word *conquistador*, which meant "conqueror." The conquerors needed workers to make the land productive. They were convinced that they were bringing a superior culture to backward people.

Las Casas did not just believe it was wrong to enslave people. He also believed it was wrong to force people to give up their own religion and customs and government.

Now it happened that Las Casas was a thoughtful man. He read books and he looked around and thought about what he saw. It may have bothered him that the Spaniards, his people, were enslaving the Native Americans. But since everyone was doing it, he did it too.

In 1512 Las Casas became a priest, but that didn't seem to have changed his life much. He was still a landowner, and he took part with other Spaniards in the bloody conquest of Cuba. He took more Indians as slaves.

About this time some priests on the Spanish islands spoke out. They said that keeping slaves was unjust. Las Casas listened, and may have been bothered, but he kept his slaves. Then one day he read these words in his Bible: "The gifts of unjust men are not accepted [by God]." That convinced Las Casas. He realized he was being unjust when he enslaved others. He reached a decision. On August 15, 1514, he preached a famous sermon announcing that he was giving up his slaves. He spoke of the injustice of slavery.

His fellow colonists were shocked and outraged. But, from that day, Las Casas never changed his ideas. He was 40 years old and lived to be 92. He spent the rest of his life traveling back and forth between Spain and the New World pleading for fairness for the Native Americans. He was very convincing. Besides, most Spaniards wanted to do what was right. Las Casas persuaded King Charles I to let him found "towns of free Indians" where Spaniards and Native Americans would work together to form an American civilization.

In December of 1520 Las Casas and a group of Spanish farmers ailed for what is today Venezuela. They started a new kind of colony 1 this New World: a colony where people of different races and back-rounds could live together in harmony. But good ideas don't always vork out. Las Casas had not planned as carefully as he might have. esides, the Native Americans just weren't interested in farming to-ether with the Spaniards. And the other Spaniards, who were living 1 the region and using slaves, thought Las Casas was a troublemak-r. They were no help. Then Indians (who may have thought all Euro-eans were alike) attacked the settlement. That was the end of that xperiment.

Bartolomé de Las Casas was not one to give up. He kept working for air treatment for the Native Americans. He wrote books. He advised

In a book called *The Tears of the Indians*, Las Casas said: "Once it happened that [the Spaniards] used 800 of the Indians instead of a team to draw their carriages, as if they had been mere beasts."

the king. He helped write laws. No one tried to keep him from speaking and writing the truth about what was happening in America.

Las Casas was convincing. But it happened that another man, who was also convincing, opposed him. That man, a scholar named Juan Ginés de Sepulveda (hwahn heen-EZ day sep-ool-VAY-dah), had a theory. It was based on the ideas of a Greek thinker named Aristotle (who had been dead for almost 2,000 years). Aristotle had said that it was "natural" for some people to be masters and some to be slaves. That seemed like a fine theory, especially to those who were masters. But, to be fair about it, in those days before modern machinery, when there was much backbreaking work that no one wanted to do, slavery seemed logical to many thinking people.

It didn't seem logical to Las Casas; it seemed evil. He said that God would destroy Spain because of the way the Spaniards were treating the Native Americans.

People in Spain were confused. They wanted to decide who was right: Las Casas or Sepulveda. They called a meeting so that both men could present their arguments.

Sepulveda said that Indians were "inferior to the Spaniards just as children are to adults, women to men, and indeed, one might even say, as apes are to men."

Many people were convinced by the learned Sepulveda, but not Las Casas. He said: "The reason why the Christians have killed and destroyed such an infinite number of souls is that they have been moved by their wish for gold and their desire to enrich themselves."

Each man spoke so well that the listeners were confused. Some people agreed with Sepulveda, and some with Las Casas. No one seemed to have won the argument, and no official decision was made.

That clash of ideas was enormously interesting to people in Spain. But in the New World it was more than just interesting: it was a matter of life and death. Indian men, women, and children were being used as slaves in the New World. Africans were being enslaved, too.

It wasn't just Aristotle who seemed to encourage slavery. People found words in the Bible and the Koran (the Muslim holy book) that seemed to make slavery acceptable. You had to really think for yourself to question those ideas. Which is just what Spain's King Charles I did. As you know, Charles ruled that Indians were not to be treated as slaves. But the king's laws didn't have much effect in the New World.

The Spaniards made Indians mine gold and silver for them. In 1510 one Franciscan friar said, "By what right do you make them die? Mining gold for you in your mines or working for you in your fields?... They lived in peace in this land before you came, in peace in their own homes."

Perhaps King Charles didn't really try hard to make them effective. Thanks to the labor of the enslaved people, gold, silver, and sugar were pouring into Spain from America. They were making the nation rich. Las Casas was spoiling things. He was like a conscience that bothers you just when you are having fun. It was greed against goodness, and goodness was losing out.

Sepulveda, who had never been to the New World, had a message that suited most Spaniards. "The Indians are like children who need someone to tell them what to do," they said. Las Casas, the good priest, was a nuisance. "He isn't realistic," they added.

The criticism didn't stop him. He kept writing. He wanted to make the record clear. He wanted to describe the Spanish New World exactly as he knew it to be. He was still writing when he was 90. The King ordered that all his writings be collected and preserved, even though most Spaniards had stopped reading his words.

Later, when the English started thinking about settling in North America, they found a book by Bartolomé de Las Casas that had been translated into English. It said, "To these quiet lands, endowed with such blessed qualities, came the Spaniards like the most cruel tigers, wolves, and lions, enraged with a sharp and tedious hunger." The English people were aghast at Las Casas's descriptions of the way the Spaniards treated the Indians. Las Casas told of an Indian girl who found a huge chunk of gold. Spanish miners, who used her as a slave, were so excited by that gold that they threw a big party and roasted a whole pig. The slave girl, who was starving, didn't even get a bite.

"We need to protect the Indians from the cruel Spaniards," the English said. "We'll treat them differently," they added. (Do you think they did?)

In the 20th century people began reading Bartolomé de Las Casas once again. Some asked themselves what the Americas would be like today if his small colony on the Venezuelan coast had prospered and grown.

The Indians died very quickly from Old World diseases. Soon Spaniards were bringing black slaves to America to work on their plantations—growing crops, such as tobacco and sugarcane. This is a sugarcane farm.

Endowed means "given" or "presented with."
Tedious usually means "boring"; here it means "very wearisome."
Aghast means "shocked" or "horrified."

30 The Big Picture

Columbus reached this island, Hispaniola, in 1492. Soon Spain had claimed half a continent.

Las Casas was a good man, and so were many other priests. Some of them devoted their lives to teaching and caring for Indians. But not all were good. Some priests were selfish, and some conquistadors were kind. History isn't easy, because people don't always behave the way you expect them to.

When you look at the big picture, what the Spaniards did is astounding. In thirty years Spain conquered more territory than the Romans had conquered in 500 years. Spanish conquistadors cut their way through steaming jungles and thick forests. They scaled mountains, crossed deserts, and defeated brave and determined Indian foes. They created new nations, a new people, and a new way of life. In Mexico and South America the Spanish built great cities. If they had found gold in North America, they would have built great cities here. And if they had, everyone in the United States might be speaking Spanish today.

The energy of the Spaniards was awesome. In Mexico and South America, where they found the gold and silver that was so important to them, they gathered the wealth of a continent and shipped it away. Other nations didn't think Spain was doing wrong. They were jealous that they weren't gathering riches, too.

Remember Martin Luther? In 1517 he nailed his famous list of complaints to a church door in Germany and began the movement that became Protestantism.

Ferdinand and Isabella tried to get rid of all the Jews, Muslims, and Protestants in Spain.

Although they didn't know it, all that gold would actually mess up the Spanish economy. Easy riches often do that. Spanish industry declined, inflation set in, taxes went up, and peasants left for America. Spain fought expensive wars with France. In addition, the Spanish Inquisition (in-kwih-ZISH-un) forced talented Jews, Muslims, and religious free-thinkers to leave the country.

The Inquisition was a Catholic court where people were brought to trial for their religious beliefs. The Inquisition came to Spain at the end of the 15th century at the request of King Ferdinand and Queen Isabella. In 1492 Spanish Jews were given a choice: they could become Catholic or leave the country. If they converted to Catholicism, but were not seen to be true in their belief, they were tried by the Inquisition and burned at the stake. The inquisitors went wild torturing and killing. The Pope tried to control them, but he couldn't do it. In 1502 the Spanish Inquisition turned on the Muslims.

A Cuban Indian chief named Hatuey tried to flee from the Spaniards. When he was about to be burned at the stake, a friar begged him to become a Christian in order to save his soul. Hatuey said he would rather go to hell than convert.

Then it became the turn of Spain's Protestants (there weren't many of them), and then the Inquisition began to attack Catholics who spoke out. No one dared protest. Some people seemed to think the Inquisition was doing the right thing. The Inquisition spread to Peru and Mexico (the Inquisition didn't end in Spain until 1834—over 300 years later).

Spain was at the beginning of a long decline. But if, in the 16th century, you had predicted that decline, most people would have laughed at you. Spain was still the most powerful nation in Europe.

In the New World she had a 100-year head start on the other European nations.

A **freethinker** is just that: somebody who thinks freely about what he or she believes, instead of just sticking to what's been told. Another word for such people was *heretic*, and it was a dangerous thing to be at this time.

31 From Spain to England to France

Mary I of England persecuted the Protestant church her father had founded.

By the middle of the 16th century everything seemed to be going right for Spain. King Philip II, who was on the Spanish throne, was married to Queen Mary I of England. That marriage gave Spain power in England. Philip's half sister, Margaret of Parma, was ruler of the Netherlands—more influence for Spain.

Spanish ships were said to be the best in the world. Spanish colonies were sending shiploads of gold and silver home from Mexico and Peru. Then King Philip invaded Portugal and captured the Portuguese throne.

As you might guess, some of the other European nations were jealous. They didn't like Spain having all that power and wealth. There was something else that made some other European nations unhappy with Spain. That had to do with religion. Spain was a Catholic nation, but the new Protestant religions were growing in other parts of Europe.

When Henry VIII was King of England, he became angry with the Catholic church. He set up a new Protestant church and made himself its leader. (That was in 1534.) He called it the Anglican church, or the Church of England, and insisted it was the "true" church. Many English men and women became Anglicans, but Henry's older daughter, Mary, was not one of them. She never liked the Anglican church, so she stayed a Roman Catholic. She was the Queen Mary who married King Philip II of Spain.

Mary tried to make England Catholic again. She had Protestants killed, and her husband, Philip of Spain, encouraged her. That didn't make them popular. The English people called her "Bloody Mary." They called him worse names.

When Mary died in 1558, bonfires were lit all over London in celebration. Then Mary's half sister, Elizabeth, became Queen—and just about everyone really celebrated. Queen Elizabeth was an Anglican. (Of course, the English Catholics weren't so happy. Now *they* were persecuted.)

In France the situation was even worse. During the last half of the 16th century, from 1562 to 1598, the French people fought eight ferocious civil wars—and all over religion. Some French people wanted to be Catholic, some Protestant, and they couldn't seem to agree to live peacefully together. Can you understand why, during almost 40 years of war, many French men and women were eager to head for the New World?

Looking back, today, it seems as if people and nations were acting just like silly little kids. Each one was saying, "My religion is better than yours." Actually, each believed God was on his side.

French Protestants flee as Catholic soldiers interrupt their church service. Do you see the soldier who is stealing from the Huguenots' poor box? (It says *Donnez aux povres*—"give to the poor.")

32 France in America: Pirates and Adventurers

Among the many things the French had never seen when they arrived on the North American coast were dugout canoes, made by burning out the inside of a big tree trunk.

The entrance to New York Harbor is called the Verrazano Narrows. Today a huge suspension bridge, also named for the explorer, spans the narrows and links Staten Island and Brooklyn.

A *privateer* is a pirate ship in a nation's service. Pirate ships were sailed by outlaws. They captured, stole, and plundered other ships and split the loot among themselves. Privateers stole too, but they were backed by a king. Privateers split their haul with the royal treasury.

As you know, it was an Italian, Cristoforo Colombo (Christopher Columbus), who first sailed to America for Spain. And it was an Italian, Giovanni Caboto (John Cabot), who first sailed for England. Does that seem strange? Italians were such good sailors that other nations hired them.

So it shouldn't surprise you to learn that when the French began to dream of territory and riches, they found an Italian, Giovanni da Verrazzano (jo-VAH-nee dah vay-rat-ZAH-no), to sail for them.

The King of France sent Verrazzano off to find a river passage through the American continent to the Orient. Everyone was sure there was one. They called it the Northwest Passage. The nation that controlled that passageway would control the route to China and its spices and jewels.

In 1524 Verrazzano sailed up the North American coast and into what would someday be New York City's harbor. No passageway was there. So he sailed on, still farther north, to Newfoundland in Canada. "My intention was to reach Cathay," he reported to the French King, "not expecting to find such an obstacle of new land as I found; and if for some reason I expected to find it, I thought it to be not without some strait to penetrate to the Eastern Ocean."

He didn't find the passageway, or gold, because there was none to be found. Verrazzano tried to tell that to King Francis I, but the King wasn't about to give up. A French privateer had captured a rich haul

126

of Spanish treasure—a ship packed by Cortés himself—and the French king was determined to find his own treasure.

He next sent a Frenchman, Jacques Cartier (jhak kar-tee-AY), to the New World. Cartier made three voyages, explored the north country—New Brunswick and Newfoundland—and brought back samples of a stone he thought was gold. It turned out to be something called "fool's gold," or iron pyrites (puh-RY-teez). The French just weren't any good at finding real gold.

The Spaniards, on the other hand, were very good at it. They had captured the Incas' gold and silver mines and were filling ship after ship with the glittering rocks. To get that treasure to Spain, ships

sailed through the Straits of Florida and then through the narrow channel between the Bahama Islands and the coast of Florida. That passageway led to favorable winds and to the best route across the Atlantic. The best route was on the Gulf Stream, which is actually a river in the ocean. Sailing on the stream is like stepping on a moving belt. Ships get a free ride on the current.

Now imagine that you are a pirate or a privateer and you want to capture a gold-filled ship. Where will you go? Certainly not to the Caribbean Sea, where you can easily be trapped and captured yourself. The coast of Florida is the perfect place. There you can seize a treasure ship and then head out to sea. In the 16th century a lot of pirates thought that way. There were pirate bases on the coast from Florida to Virginia. Because of that, the Spaniards needed their own Florida base to protect their ships. (You'll soon see that they established one.)

The French were interested in piracy, too. If France had a base on the Florida coast near the Gulf Stream, it just might be a good spot for French ships to pick off Spanish treasure galleons. Besides, the French still intended to search for gold in North America, and to look for those seven cities. (Yes, Cíbola again!) So on April 30, 1562, two small ships with 150 Frenchmen aboard landed in Florida (at a site that later became a Spanish town called St. Augustine). Their leader was a remarkable man named Jean Ribaut (jhon ree-BOW); an expert sailor, a devout Protestant, and a patriotic Frenchman. The story of his trip, *The Whole and True Discovery of the Land of Florida*, was published a year later.

The Indians of Florida greeted the Huguenots with friendliness. They offered the French captain clothes made of furs and "a large skin decorated with pictures of wild animals."

A *galleon* is a big, three-masted sailing vessel. Spanish ships were spacious and seaworthy, but they weren't as fast or agile as the small pirate ships.

127

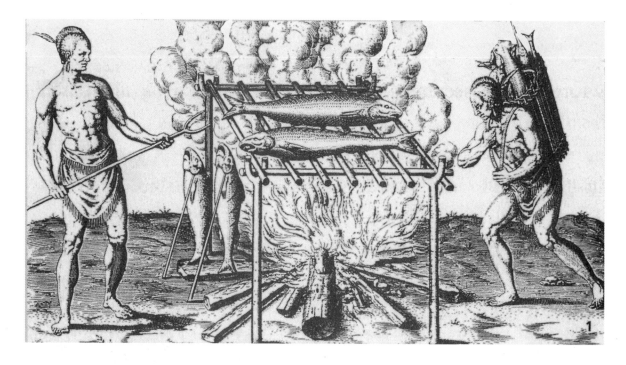

The Indians of the East Coast were experts at smoking fish. The barbecue had to be high enough so the fish did not burn or the wood grill catch fire.

In 1598 Henry IV of France gave religious freedom to Protestants with a ruling called the Edict of Nantes. In 1685 King Louis XIV revoked (took back) the Edict and persecution began again. The French Protestants headed for the New World once more.

Protestants, known as Huguenots (HUE-guh-noes), were being persecuted in France because of their religion. Ribaut believed this colony would give the Huguenots a chance to prove their loyalty to their country, to be heroic, and to gain land and gold for France.

Things began well. Ribaut wrote that the Indians were "very gentle, courteous and of good nature....They were all naked and of good stature." The country was "the fairest, fruitfullest and pleasantest of all the world. The sight of the fair meadows is a pleasure not able to be expressed with the tongue."

Ribaut's group made its way north to what is now the coast of South Carolina. There they built a fort called Charlesfort. But they were soon in need of food and ammunition and farming tools. They were sailors and hadn't planned to grow crops. Ribaut decided to go back to France for supplies; thirty of his men volunteered to stay in America.

Back in Europe, things were all upset—from those religious wars—so it wasn't easy for Ribaut to return with supplies. In the meantime, the Frenchmen at Charlesfort were in trouble. They had been depending on the Indians for food, but the Indians stopped feeding them. That was because the Frenchmen had taken part in some fights between Indian tribes—and double-crossed both sides. They ended up making everyone mad at them.

The situation was bleak. The Huguenots were starving, and suffering from New World germs. In addition, they were probably homesick. So they built a boat, used their shirts for sails, and set out for France. At sea they really did starve. They ate their shoes, and—I'm sorry to tell you—one of the sailors. They were drifting, half dead, when an English ship found them and took them to London.

In England everyone wanted to meet them and hear their tale, even Queen Elizabeth. The French sailors raved of the beauty and wonder of America. (When they were there, they couldn't wait to leave!) Their stories of the New World got fancier and fancier. One spoke of precious jewels that could be mined only at night because in daylight they blinded men with their dazzling reflections. All told of gold and silver, and of pearls and spices.

Ribaut, who had been fighting in France (with the Huguenots against the Catholics), turned up in England, where he also met the Queen. The English were soon reading Ribaut's book. It was 1563, and the British caught New World fever. They began making their own plans for settlements.

One 16-year-old French boy at Charlesfort thought the plan to sail home was crazy. He chose to live with the Indians and was later picked up by a Spanish ship.

An Abundance of All That Man Might Desire

Nicolas Le Challeux (shall-UH) sailed from France with Captain Ribaut in 1565 and wrote about his adventures:

...the King, with the princes and nobles of the court, decided to send a large number of men and many ships to one of the countries of India, called Florida—a land recently discovered by the French.✶...Jean Ribaut, a man of courage and wisdom, and well-experienced in navigation, was summoned to the court, where he received the King's commission to outfit seven ships to transport men, food, and arms....

The news of this voyage was quickly spread abroad, and many men were persuaded to serve under the command of this captain and the King's authority. They were moved by various reasons: some enlisted with a sincere desire to know and see the country, hoping the mission would profit them later. Others wanted with heart and soul to make war, preferring to risk the rage of the waters rather than remain in their accustomed condition—which was just as precarious. ✶✶

The rumor spread here that Florida promised an abundance of all that man might desire in the world...there was neither frost nor snow there, nor any northern cold, and it escaped the burning heat of the South. Without labor or tillage, the ground brought forth enough to sustain the life of the natives, as well as of those who come to dwell there....

The country is also rich in gold and in all sorts of animals, both tame and wild....There are high hills, pleasant streams and rivers, and various kinds of trees fill the air with sweet scent....

Men were persuaded by these promises, or by covetousness (believing they would be made rich by this voyage because of the gold), and came in legions to the town where the muster for the voyage was held.

✶ Florida, a country of India? Discovered by the French? Who wrote this story?
✶✶ If they were Protestants, they were in trouble in France; they might as well go to sea. What does *precarious* mean? How about *covetousness*?

33 Rain, Ambush, and Murder

Pirates and privateers lurked along the coast, waiting for galleons.

The French weren't finished in Florida. In 1564 they returned to America and built Fort Caroline (near today's city of Jacksonville). The story of what happened to that fort is a story of politics and pirates, of Indians and murder. The main characters are Ribaut and Pedro Menendez de Aviles (PAY-droe men-EN-dez day ah-VEE-lace), one of Spain's top naval officers.

Menendez set sail from Spain in June of 1565 with 1,500 Spaniards, along with breeding animals, seeds, and farming tools. He was to do three things: get rid of the French, build a fort to protect the Spanish fleet, and do some exploring in La Florida. (To the Spanish that meant North America.)

Menendez had a vision of La Florida as a great Spanish colony, with an economy based on pearl fishing, agriculture, and mining. But first he had to do something about the French at Fort Caroline. He established a Spanish base at St. Augustine.

It was hurricane season and raining ferociously, but that didn't stop Menendez. He and his men marched for three days northward from St. Augustine, through swamps, to Fort Caroline. The Spaniards were often in water up to their waists. They shared the swamps with poisonous snakes, enormous spiders, and ravenous mosquitoes.

The French in their fort thought they were safe. They couldn't imagine anyone out in that terrible downpour. "The rains continued as constant and heavy as if the world was again to be overwhelmed with a flood," wrote a Frenchman who was there.

So no one was on guard when the Spaniards attacked. Too bad for the French; the battle didn't even last an hour. When it was over, most of the French soldiers were dead. Some who surrendered were murdered.

Safe Home

Two of those who made it back to France from Fort Caroline were men we have heard of already: an artist, Jacques LeMoyne, and a carpenter, Nicolas Le Challeux. Le Challeux wrote a book full of details about Indian life and the adventures of the French in America (you read a little of it in the previous chapter). Engravings from LeMoyne's paintings (remember the Indian deer hunters?) were used to illustrate the book.

A few made it to a ship and back to France, where they told of the awful massacre.

In the meantime, Jean Ribaut didn't know about the attack. He had sailed south from Fort Caroline intending to destroy the Spaniards at St. Augustine. But the French had run out of luck. A hurricane smashed the Florida coast, and Ribaut's ships were wrecked and sank. Ribaut and most of his men made it to an island where they were marooned with no food and no way to leave. There was nothing they could do but surrender to Menendez. He murdered them all, except ten who were Catholics. Once again Spain controlled the Florida coast and its important shipping lanes.

But that is not quite the end. The Spaniards built San Mateo where Fort Caroline had stood. They built Santa Elena at today's Parris Island in South Carolina. They built a series of small forts running up the coast from St. Augustine to Santa Elena. They searched for gold in North Carolina. And they sent five priests and four helpers to the Chesapeake Bay area to establish a settlement.

This is what happened next:

The French sent a revenge mission that wiped out San Mateo in 1568.

Indians murdered the five priests and three of their helpers, but they spared 13-year-old Alonzo de Olmos. Menendez sailed to Chesapeake Bay to check on the settlement, found Alonzo, and learned of the murdered priests. He captured eight Indians and hanged them from the yardarm of his ship. It was 1571.

England's swashbuckling sea captain, Francis Drake, attacked Santa Elena in 1586. The Spanish settlers fled. Then Drake

By 1600 the Spaniards and French had several forts on North America's east coast. Pirates like the infamous Blackbeard preyed on their ships.

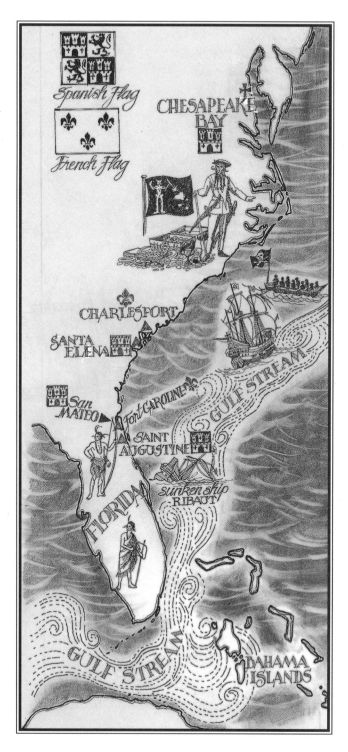

The **yardarm** is the cross-piece on the mast of a square-rigged ship. When the sun is over the yardarm, it's after noon. **Swashbuckling** means being bold, taking risks, and showing off a bit.

burned and plundered St. Augustine. But it survived to become the first permanent European settlement on the North American continent.

For the next century and more, France, England, and Spain argued and fought over who had the right to settle the land the English called Virginia, the Spaniards called Florida, and the French called New France. No one asked the Native Americans for their thoughts on the matter.

Drake's ship, the *Golden Hind*, attacks a Spanish treasure galleon, *Nuestra Señora de la Concepción,* laden with gold.

Who Am I?

A long time ago, in the days before history books, when just a few people roamed the earth, and they lived in caves, I was big. Some-

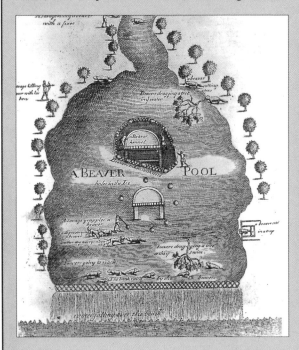

times I grew to weigh 700 or 800 pounds. Then I slimmed way down. Today I usually weigh about 60 pounds when I'm fully grown. (That is about the same weight as some 10-year-old humans.)

When the first Europeans arrived in the New World there were about 60 million of my kind, living from Mexico to the Arctic Circle—four times more than all the people today living in Los Angeles, New York, Chicago, Boston, and Washington, D.C. The Europeans turned us into hats, robes, and blankets. By 1895, only five of us were known to exist in New York State. Then people brought some of my cousins from the Far West to New York. I soon began multiplying. (Not doing arithmetic, silly, but increasing my family.)

Have you figured out who I am yet? More clues: I am a superbuilder. I cut down trees, eat the bark for dinner, and build dams with the logs. Some of those dams are more than 1,000 feet long. The dams create ponds behind them. After I leave a pond it often fills up with mud and becomes a meadow. Like man, I change my environment. I usually make it better. I thin trees, control erosion (that's when the land washes away), and keep water levels high. All that helps prevent forest fires.

How about you? Do you have any ideas about what to do to make the world better?

34 New France

Like buffalo, so many beavers were killed after the Europeans came that in some places they nearly died out.

After the massacre in Florida the French moved far north, to the region that is now Canada. There they stuck to fishing, fur trapping, and trading. Frenchmen had been doing that in America since the beginning of the 16th century.

The French found the waters near Newfoundland so thick with fish that boats had a hard time passing through. They came in the spring, fished all summer, salted and dried their catch, and went home for the winter.

Sometimes strange things influence history—sometimes even hats. People in France and England were crazy about beaver hats. They were the latest fashion and the American woods were full of beavers. So Frenchmen came to America to trap beavers or to trade with the Indians for beaver skins.

Those who came usually got along with the Indians. They had to—there were too few Frenchmen to start a fight.

The French fishermen and fur traders lived much as the Indians did. They became friends with the Algonquin and Huron tribes and traded with and learned from them.

The Algonquins and Hurons were enemies of the powerful Iroquois Indians.

For trifles, as knives, bells, looking glasses, and such small merchandise which cost him four English pounds, a French trader had commodities that sold at his return for 110 pounds.

—RICHARD HAKLUYT, 1598

After his first visit to the New World, Samuel de Champlain wrote a book about what he had seen. He drew pictures and made good maps, too.

The English have no sense; they give us twenty knives...for one beaver skin.

—AN ALGONQUIN INDIAN, 1634

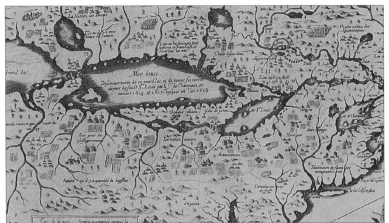

This is Champlain's map of some of his explorations in "Nouvelle France" (New France). In the middle are some of the Great Lakes—can you tell which they are? The smaller lake near the coast on the right was given its name in the 17th century. It still has the same name today—Lake Champlain.

Robust (roe-BUST) means strong, healthy, and sturdy.

So the French took sides with their friends against the Iroquois. A French explorer, Samuel de Champlain, joined the Algonquins and Hurons in a raid on their enemy. The French brought their arquebuses (ARK-wuh-buses), big clumsy weapons. The Indians called them "thunderhorns." Never before had the Iroquois seen guns. Three Iroquois chiefs were killed. A hundred years later the Iroquois would still hate the French because of the deaths of those chiefs.

It's too bad Champlain made that mistake, because everything

else he did was outstanding. He was not only a fine sailor and leader, but also a writer, artist, and mapmaker. Writing a description of what he thought "a good captain" should be, he didn't realize he was describing himself. "An upright, God-fearing man, not dainty about food or drink, robust and alert, with good sea legs."

Samuel de Champlain dreamed of a new French nation in America. He realized that if the French wanted to stay in the New World, they would need to bring their families and start colonies where French people could live.

In 1608 Champlain founded a settlement called Quebec. Today Quebec is the second-oldest continuously occupied city on the North American continent. You already know the oldest city. (If you don't, reread the last chapter! You know city number three, too. Think West. Have you got it?)

Notice those words, "continuously occupied," meaning cities that still exist. Other early settlements disappeared.

Samuel de Champlain understood that the American continent was a wonderful place to live, but few people in France agreed. Life seemed good enough in France; there was no need to leave—especially for an unknown wilderness. The French Huguenots, who were persecuted because of their Protestant religion, might have come, but their faith made them unwelcome in Catholic Canada. So New France grew, but slowly. Brave and energetic explorers helped it grow.

Champlain helped Huron and Algonquin Indians fight the Iroquois. In this battle in 1609 a few of the French and 60 Hurons surprised 200 Iroquois and beat them.

Traveling by Canoe and Portage

Do you see that Catholic priest over there—the Frenchman in a black cape? He is Jacques Marquette (jhak mar-KET). He came to Canada just six years ago and already knows many Indian languages. Most people here call him "Père (PEAR) Jacques." *Père* is the French word for "father"; a priest is a Catholic "father." Marquette is a gentle, sweet man, and nothing seems to disturb him—not even the worst hazards of the wilderness. He is a Jesuit (JEZ-yoo-it), a member of a scholarly order of priests who have come to New France to teach the Indians about Christianity. Marquette goes freely among the Indians; they love him. Native Americans call the priests "black robes," because of the long, dark garments they wear.

We have come here in our time capsule. Notice: it is 1673. (Just for this chapter, we are leaping ahead to the 17th century. That is to give you an idea of what the French explorers will accomplish in this New World. In the next chapter we will be back in the 16th century.)

Repeat: it is 1673. The governor of New France, Louis Frontenac, has ordered Father Marquette and Louis Joliet (LOO-ee joe-lee-AY) to find the Northwest Passage. For almost 200 years, Europeans have searched for that route through America. The French are determined to find it. Like Columbus, they want to sail west and reach the Far East.

Joliet is a mapmaker. He was born in New France; a priest says he has "the courage to fear nothing where all is to be feared."

Perhaps this team will find the Northwest Passage. Marquette and Joliet go in birchbark canoes with five other Frenchmen and two Indian guides.

Let's follow them. We will be gone four months and will travel 2,500 miles. Marquette and Joliet will take notes and make maps as they go. Mostly they will eat "Indian corn, with some smoked meat." (Father Marquette writes this in his journal.) We will watch as the French explorers paddle on the Wisconsin, Mississippi, Illinois, and Chicago rivers. Sometimes it will be necessary to carry the canoes overland from one river to another. This is called portaging.

Watch out! Indians are attacking. Will we live through the raid?

Don't worry. Father Marquette will save everyone. The Indians see the peace pipe he smokes. The Illinois Indians gave it to him, and other tribes respect it—and him.

This expedition makes an important discovery. Marquette and Joliet learn that the Mississippi River empties into the Gulf of Mexico, not the Gulf of California, as had been thought. But this expedition will not get to the river mouth. The Indians tell of Spaniards with guns; it is wise to head back to Quebec. Marquette and Joliet have not found the Northwest Passage (and it will not be found until the 20th century; the passage is so far north that most of the year it is under thick ice).

But they have seen wondrous things. In Illinois, Marquette reports, they "saw two monsters painted on...rocks, which startled us at first, and on which the boldest Indian dares not gaze long. They are as large as a calf, with horns on the head like a deer, a fearful look, red eyes, bearded like a tiger, the face somewhat like a man's, the body covered with scales, and the tail so long that it twice makes the turn of the body...these two monsters are so well painted...good painters in France would find it hard to do as well."

Marquette and Joliet claim much land for France. That the Indians have lived here for centuries does not seem to concern France. Nor does it concern England or Spain or any of the nations that attempt to colonize America. In Paris, French leaders decide to let England have the eastern edge of North America (below Nova Scotia); France will take the rich interior lands.

If we zoom on, nine years in time, we will meet René Robert Cavelier (ruh-NAY roe-BEAR kav-ul-YAY), a Frenchman with the title of Sieur de La Salle (syur duh la SAHL). A French sieur is like an English lord or knight. Robert Cavelier is Lord La Salle. He has received a grant of New World land from the king; he will sell it to pay for this expedition. That is how important it is to him to be able to explore.

La Salle is fearless and stubborn. He will travel even farther and faster than Marquette and Joliet. He is told the Mississippi is "guarded by monsters, tritons, giant lizards, and barbarous nations who eat their enemies." He believes that may be true. Besides, he knows that Spaniards are in the Gulf of Mexico, and they are as dangerous to him as any monster.

But La Salle isn't the kind to be stopped by a few monsters—or Spaniards. He hikes and canoes from Canada down the Mississippi to its mouth at the Gulf of Mexico, and then claims the river and everything west of it for France and the French king, Louis XIV. La Salle names the territory Louisiana in honor of the king.

After that he heads for France to tell King Louis of his adventures. But La Salle hasn't finished exploring. He sails back from France to the Gulf of Mexico. He is trying to find the Mississippi River by coming from the other direction. He can't find it. He goes too far, lands in

La Salle (that's him on the opposite page) became the first European to explore the entire Mississippi River (the thick black line running all the way down near the middle of the map).

Texas, and marches his men so hard and fast that most of them die. Finally, those who are left can take no more. They murder La Salle and leave his body 350 miles from the Mississippi.

Because of those French explorers, France claims all of

Canada and all the land west of the Mississippi. It is a vast territory—the French have no idea of its actual size. They hope that it contains riches. They are eager to find wealth on this great continent, to "civilize" its peoples, and to bring them Christianity.

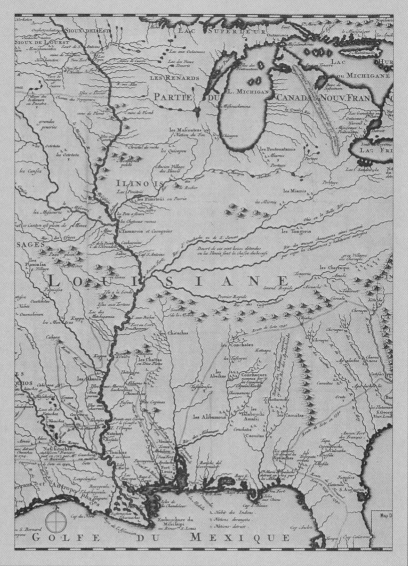

35 Elizabeth and Friends

The word **court** has many different meanings. There is a basketball court, a law court, and a royal court. A royal court usually consists of a king or queen and his or her advisers and servants. It can also be the king and queen's home.

One of the plays that William Shakespeare (above) wrote was *The Tempest*. Caliban, a character in the play, was a kind of monster, based on explorers' descriptions of the Patagonians.

Shakespeare's

Globe Theatre is actually shaped like a polygon. It still stands in England (it's been rebuilt), where you can visit it. Did you notice the *re* on the end of *theater*? That is the way the word is spelled in England, where they pronounce some words strangely, too. (Of course, they think that we are the ones who spell and pronounce words strangely.)

And now, back to the late 16th century—to the time of Menendez and Ribaut, to the Elizabethan Age.

As you know, Elizabeth became queen of England in 1558, when Bloody Mary died. She turned out to be amazing, perhaps the best monarch England ever had. She was smart, tough, and energetic. She was a musician and a poet. She spoke French, Spanish, and Italian, and could read and write Greek and Latin. She filled her court with people who were intelligent and witty and got things done. The Elizabethan Age was the greatest of times in England.

It was then that Shakespeare wrote his plays and English developed into a language of richness and beauty. It was an age when people wanted to act like knights and be chivalrous. Sir Philip Sidney was an Elizabethan poet and a knight whom everyone admired. When he was wounded on a battlefield, he passed his cup of water to a dying soldier, saying, "Thy need is yet greater than mine."

Sir Walter Raleigh was another of the great men of his day. He wrote poetry, fought pirates, and had adventures. One day, while Queen Elizabeth was out walking, she came to a huge puddle (or so they say). There was no way around it. Sir Walter saw that she would dirty her feet, so he whipped off his cloak, spread it out over the mud, and let her walk across it. That was chivalry! (Even if it wasn't true.)

Good manners make you feel good about yourself. Most English people felt very good about themselves and their country during the Elizabethan Age. For those with ambition, energy, and luck, there were new opportunities.

For some, life was becoming elegant. Horse-drawn carriages were introduced into England from the Netherlands. Forks were being used at the French court, and a few Englishmen actually tried them. (But most people still ate with only a knife and their fingers.) Before the end of the century some people would put heels on their shoes. And picture this marvel of technology: a toilet, called a water closet, was installed in the queen's palace.

The Elizabethans ate heartily. Those who could afford them downed meat, fish, bread, wine, and sweets, all at the same meal. They didn't eat vegetables. Vegetables were eaten only by the poor who grew them in gardens. The rich sometimes got painful joint ailments. Vegetables might have prevented them. Many people, even Queen Elizabeth, had teeth blackened by decay. Elizabeth's teeth were painful as well as ugly. She tried to clean them the way the experts advised: by brushing them with sugar!

The Elizabethan Age was a time of excess. Some people ate too much, drank too much, and spent too much money on clothes and partying. The English people also began to create a great empire that would one day stretch around the world. The queen seemed to infect the nation with her taste and energy. That energy would change America, but not in the 16th century.

It was Spain that England had on her mind in the 16th century. Spain was a greater power than England. The English wanted their nation to be the world's greatest. The French and the Dutch wanted the same thing for their nations. Spain would fight to hold on to her exalted position. That duel for first place between the great European powers seemed much more important to most people than anything that might develop in the wilds of a distant continent.

England did try to plant colonies in the New World during Elizabeth's reign. She just didn't seem to have much luck.

Not long before the queen died, she said: "I count the glory of my crown that I have reigned with your loves...and though you have had, and may have, many mightier and wiser princes sitting in this seat, yet you never had nor shall have, any that will love you better."

36 Utopia in America

The Latin motto on Raleigh's coat of arms says "with love and courage."

When "Sir" appears before a man's name it tells you he was knighted by the English king or queen. Thomas More was a great man who was both knighted and killed by a king. It's an interesting story, but you'll have to find out about it on your own.

Barbarous (BAR-ba-russ) means wild, primitive, harsh, or cruel.

Picture a long, stout rope. On each end of the rope, strong teams pull hard. Sometimes the rope is pulled in one direction, sometimes in the other. Mostly, however, the teams are even. They balance each other in a kind of tension.

And so it was, and is, and always has been in North America. From the beginning, the Europeans who came to America had two dreams:

There was *the dream of riches*, of America as a land of gold where one could become wealthy.

And there was *the dream of a new world*, of an ideal place where the mistakes of Europe could be avoided, where people could pursue happiness. Sometimes those dreams pulled in opposite directions; sometimes they worked in harmony.

Sir Humphrey Gilbert may have held both dreams. He was the first Englishman to hold a royal charter to "have, hold, occupy and enjoy...remote heathen and barbarous lands." Those barbarous lands were America; Sir Humphrey meant to settle English people here. His charter said that all settlers

Raleigh and his son. How would you like to play in those clothes?

140

should "enjoy all the privileges of free... persons native of England" and that all laws should be as close to English law as possible.

Now that was unusual. English men and women would lose no rights when they moved to the new land. They would be entitled to trial by jury and other English rights. The head of the colony (sometimes called the *proprietor*, which means the owner) could not be a dictator.

Sir Humphrey Gilbert never got a chance to use his charter: he was lost at sea. His small ship, the *Squirrel*, was swallowed by a huge wave. Just before he went under, Gilbert was seen reading a book called *Utopia*, written by Sir Thomas More.

More's book was written as a kind of a joke—with a serious thought behind it. *Utopia* was about an island. Sir Thomas described its people, its government, and its way of life as being close to perfect. English people who could read Greek knew that the word *utopia* actually meant "no place," and that Utopia didn't exist. But only a few of More's readers understood Greek. Those who could read Greek also knew that the name of the book's sailor hero, Raphael Hytholadaeus (ra-fee-YEL hy-thuh-la-DAY-uss), meant "skilled in talking nonsense."

But many people who read More's *Utopia* thought it a description of a real place. One missionary made plans to go there to convert the inhabitants to Christianity.

Sir Humphrey Gilbert could read Greek, and he knew there was no real Utopia. But he liked Sir Thomas More's ideas. We think he hoped to set up a utopia in America—a close-to-perfect place to live—and perhaps find gold, too.

Remember those two dreams: gold and a good society—that was what many Englishmen hoped for in America, especially Sir Walter Raleigh.

Sir Walter was Sir Humphrey's half brother. After Sir Humphrey Gilbert drowned, Sir Walter Raleigh decided to take over his charter and dream. The Queen was happy to have him do it. Raleigh was not

An Englishman brought home tobacco and told his countrymen: "The Floridians have a kind of herb dried, who with a cane and an earthen cup on the end, with fire, do suck through the cane the smoke thereof, which smoke satisfies their hunger." The Europeans learned the new technique very quickly.

Sir Walter Raleigh brought back no gold and silver from the New World. But he did bring a new fashion, copied from the Indians: tobacco smoking.

Amauroti vrbs.

Fons Anydri.   Oftium anydri

This map of imaginary Utopia was drawn to illustrate the book in 1518, when people were fascinated by the reports of the New World.

only chivalrous, he was handsome, and Queen Elizabeth liked him—a lot.

Sir Walter Raleigh sent three expeditions to the New World.

The first one went to look the place over and pick out a good spot for a colony. The ships' captains came back with a report of a wonderful land and of Indians who were "most gentle, loving and faithful...and live after the manner of the golden age." The captains were describing the coastal area that would someday become North Carolina. Raleigh named the whole land Virginia, after Elizabeth, who was called the Virgin Queen. (Queen Elizabeth never married, and an unmarried woman is sometimes called a virgin. That's how the name Virginia came about.)

Raleigh's second expedition was a big one: in 1585 seven ships sailed with 100 men. They planned to start a colony. The men included John White, an artist; Thomas Harriot, a famous mathematician, poet, and astronomer; Thomas Cavendish, a navigator who later became the third man to sail a ship around the world; and a Jewish mineral expert from Bohemia (which is now part of the Czech republic) named Joachim Ganz, who was to search for gold and silver.

Ganz found copper but no precious minerals; White drew pictures; and Harriot wrote a story of their adventure. We have the story and pictures today.

The men found that colonizing was a lot harder than they had expected. They were homesick and hungry when an English ship commanded by Sir Francis Drake came by to check on them. All climbed aboard and went home.

Raleigh tried once more. In 1587 he sent out a new colony. Its mission was to establish the city of Raleigh in Virginia.

37 Lost: A Colony

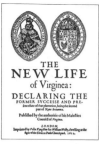

THE
NEW LIFE
of Virginea:
DECLARING THE
FORMER SVCCESSE AND PRE-
sent estate of that plantation, being the second
part of *Nova Britannia.*
Published by the authoritie of his Maiesties
Counsell of *Virginia.*

LONDON,
Imprinted by *Felix Kyngston* for *William Welby,* dwelling at the
figne of the Swan in Pauls Churchyard. 1612.

Companies put out fliers like this to attract settlers to the New World.

Sir Walter Raleigh dreamed of an English nation in America. As you know, the Queen gave him a grant to try to start one.

In books to come, you will read more about rulers giving grants to America. A grant is a deed of land. Kings and queens were giving away America. What made them think they had a right to that land?

Unfortunately, it had something to do with European arrogance. Arrogance is a strong word. The dictionary says that someone who is arrogant overestimates his importance. An arrogant person is stuck-up. An arrogant nation is likely to go to war or push other nations and peoples around. The Aztecs were arrogant when they conquered neighboring peoples and sacrificed the captives to their gods. It is too bad to have to describe Europe as arrogant. It just happens to be true.

Many Europeans in the 16th and 17th centuries saw themselves as civilized and important. They saw the Indians as savage and unimportant. To the Europeans, America was empty. It was as if the Indians didn't exist. Since the country was empty—in their eyes—it was available to anyone who could grab it. And so they raced for it; Spain, England, France, Portugal, the Netherlands, and Sweden all claimed parts of the land. Sometimes they fought with each other over those claims.

Although England was a latecomer, in 1587 it looked as if she might get going. Sir Walter Raleigh was trying again. This time he sent families to found a colony. He finally realized that you can't have a real colony without women and children.

But the settlers didn't know how to survive in the wilderness.

Samuel de Champlain arrived in New France in 1603. Do you remember when Quebec was founded?

The first three permanent colonies in the lands that would become the United States and Canada were St. Augustine, Quebec, and Santa Fe.

NOVA BRITANNIA.

OFFRING MOST

Excellent fruites by Planting in
VIRGINIA.

Exciting all such as be well affected
to further the same.

LONDON
Printed for SAMVEL MACHAM, and are to be sold at
his Shop in Pauls Church-yard, at the
Signe of the Bul-head.
1609.

We've seen "New Spain" and "New France." Can you guess what "Nova Britannia" means?

They spent too much time thinking about Spaniards and not enough time looking for food or building shelters. Some of them hoped to turn their settlement into a base for raiding Spanish ships. Most hoped to find gold. They built a fort to protect themselves from Spanish attack.

The colonists landed on Roanoke Island late in the spring of 1587—too late to plant crops. The artist John White was back, this time as leader of the group. He brought his sketch pad and paints. White's daughter, Eleanor Dare, came with him. White soon realized they would need more food; he decided to go back to England for supplies. Just before he left, Eleanor Dare gave birth to a baby. The baby, John White's granddaughter, was named Virginia. Virginia Dare was the first English baby born in the New World.

When White returned to England, he found the country was fighting Spain. The Queen would not give him a ship to go to America because she needed all her ships in England. One thing after another happened, and it took three years before John White

Guns vs. Arrows

When the Native Americans first met the European invaders, it was arrows against guns. Of course you know which was the superior weapon. Or do you?

One weapon had a big advantage psychologically (sy-koh-LODGE-ick-al-lee), which means "in people's minds." It was the weapon that made a big noise. If you heard a huge bang, saw smoke, and watched as someone dropped down, you'd probably do what the natives did—run.

But the Indians actually had the better weapon. Their arrows were deadly, traveled farther, and were easier to control than the bullets of that time. A skilled bowman could shoot six arrows in the time it took to shoot one bullet. Reloading a musket was a slow process.

Muskets were not very accurate. But they did make noise. And they were complicated. The bow and arrow was simple and silent. Everyone—on both sides—thought guns the superior weapon. Naturally, the Indians had to have guns.

Soon they did. In order to shoot their guns, they had to have ammunition. That made them dependent on the white men. Some Indians would do almost anything to get guns and ammunition. That meant that white men could make deals with different tribes

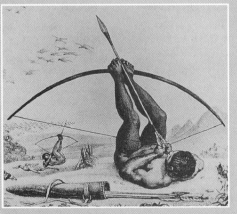

These archers shot at migrating birds. A big bow needed the strength of legs as well as arms.

and set one tribe against another. And all for muskets that were not as accurate as the bows and arrows the Indians could make easily.

What is to be learned from this?

was able to return to Roanoke Island. When he got there, the colonists had vanished. The letters "C R O" were carved on a tree. John White knew that Croatan was the name of a nearby island (and also the name of an Indian tribe). Had the colonists gone to Croatan Island? Had Indians attacked?

White was desperate. He wanted to find his daughter, his granddaughter, and the other colonists. But a treacherous storm was on the horizon. The captain and crew decided to sail away. (For John White's own words telling exactly what happened, turn the page.) Naturally, John White was very, very upset. So were other people in England. They began talking of the "Lost Colony."

In 1603 an English ship with a cargo of valuable hardwood stopped at Roanoke Island before heading back to England. Six men rowed ashore to explore and see if they could find the lost colonists. They didn't notice the Indian warriors hiding behind the trees. The Indians killed five of the men; one survived and wrote their story.

Many ships were wrecked off the long, skinny islands (the Outer Banks). Protected behind the Outer Banks is Roanoke Island. Can you see the barricaded village and the fish trap in the water?

Treacherous (TRETCH-ur-uss) means unreliable or untrustworthy.
The **horizon** (hur-IZE-un) is the farthest point of the earth that a person can see. On the ocean, it is the line where the sea and the sky seem to meet.

No one has ever solved the mystery of the Lost Colony. Some Indians in North Carolina say they are descended from people with gray eyes. Indians usually have brown eyes. Could Virginia Dare have become a gray-eyed Indian?

A few years later England tried another settlement, this time at a place we know as Maine. But freezing weather got to the settlers (as well as a lack of supplies), so they sailed back to England.

If it wasn't poor planning, then it was the weather or the Spanish or the Indians. England just couldn't seem to get herself planted in American soil.

This was the fortified longhouse village of Pomeiock in Virginia. Do you see the giant sunflowers growing on the far side of the palisade?

146

From John White's Log

If John White was to get back to Roanoke Island, he had to hitch a ride. And that is just what he did. On March 20, 1590, he climbed aboard the Hopewell, *one of three ships on their way to the Americas to do some privateering. The ships headed for the Canary Islands, where they got fresh water and found favorable winds. By April 30 they were in the Caribbean Sea. They soon captured a 10-ton frigate "coming from Guatemala with a cargo of hides and ginger." It wasn't the only plunder they took.*

In mid-August they were ready to head home. They planned to stop at Roanoke Island before heading east across the Atlantic. Roanoke is inside the string of islands—called the Outer Banks—that elbow out off the coast of today's North Carolina. To reach Roanoke, the ships' captains needed to get through a breach—a gap—between the outer islands. They were in a hurry. A bad storm—a northeaster, as they are still called— was creating mountainlike waves. Here are John White's own words (words written more than 400 years ago) about what happened next:

A Spanish frigate of the kind that John White's ship, the Hopewell, *captured on its outbound trip.*

Captain Spicer and Skinner hung on until they sank and were seen no more....Four of the men who could swim...were saved by Captain Cooke. He took off his clothes and, with four others who could swim very well, rowed out as fast as possible and saved the four of them. There had been eleven men in the boat, and seven of the best were drowned....

The accident so upset the sailors that they were all of one mind not to go any farther to search for the planters. But later, they got ready the boats....

At daybreak we landed [at Roanoke], and we... proceeded to walk along the shore, rounding the northern part of the island, until we came to the place where we left our [first] colony in the year 1586....As we went inshore up the sandy bank we saw a tree on the brow of a cliff curiously carved with the clear Roman letters C R O.

We knew at once that these letters indicated the place to which the planters had gone. Before I left them we had agreed on a secret token. They were to write or carve on trees or doorposts the name of the place where they had settled....

The weather grew fouler and fouler. Our food supply was diminishing, and we had lost our cask of fresh water. We therefore decided to go...we hoped to...visit our countrymen in Virginia on the return trip.

We nearly sank while getting through, for a great sea broke into the boat and filled us half-full of water, though by the will of God and the careful steering of Captain Cooke we got safely ashore. But much of our equipment, food, matches, and powder was wet and spoiled.

[Another boat was not as lucky.] The wind was blowing a great gale from the northeast into the harbor. The breakers [waves] were very high...and the tide pulled strongly....The rash steering of Ralph Skinner, the master's mate, let a huge sea break into the boat and capsize [overturn] it. The men stayed with the boat, some in it, others clinging to it....the waves beat against it so that some of the men were forced to let go their hold and try to wade ashore. The sea beat them down again and again. They could neither stand nor swim, and the boat turned keel upward two or three times. [The keel is the spine of the boat, running along its bottom.]

John White was never again able to get back to Virginia.

38 An Armada Is a Fleet of Ships

After their first battle with the Armada, the English almost ran out of gunpowder.

An *isthmus* is a narrow neck of land joining two larger landmasses. Do you know what happened to the Isthmus of Panama?

England wanted some of that Spanish gold, and she got it the only way she could: by stealing it. Some Englishmen said that since the Spaniards had stolen the gold from the Indians, they had a right to steal it from the Spaniards. What do you think of that argument?

An Englishman named Francis Drake didn't consider arguments at all. He just went after Spanish treasure. Drake was a daring and fearless seaman. He hated the Spaniards, and they didn't like him either. Queen Elizabeth gave him a privateering commission, and he used it to raid Spanish ports, sink Spanish ships, and burn Spanish towns. On the Isthmus of Panama he captured three mule trains heaped with 30 tons of silver and loaded the loot onto his ships. Everywhere he landed, Drake struck terror and gathered riches.

In 1577 Drake sailed through the Straits of Magellan and into the Pacific Ocean. His ship, the *Golden Hind*, was the first English vessel to reach the western side of America. Spanish ships were unguarded on that western coast because they thought they were safe from privateers. Then along came Drake. Near the coast of Peru he captured a huge treasure ship full of gold. The *Golden Hind* became so heavy with gold it began to ride low in the water. That didn't stop Drake. He sailed on—to California and even farther north—trying to find the Northwest Passage from the West Coast. When he couldn't find it, he headed west, to China, picked up spices, and brought everything home to the queen. It was quite a haul for Drake, England, and the queen. The English people

Drake captured treasure worth twice the queen's income on his world voyage.

were proud: Drake was the first sea captain to take a ship around the world. (Remember, Magellan hadn't lived to finish his voyage.) The queen made Drake a knight: he was now Sir Francis Drake.

In England he was a national hero, as famous as anyone in his time. But to the Spanish ambassador he was "the master-thief of the unknown world." That should tell you something about history. One nation's hero is often another's villain.

Even though some of her treasure ships were being captured, Spain was still the greatest power in the world—at least that is what everyone in Europe thought. So, in 1588, when the Spaniards sent a great fleet of ships—called an *armada*—to fight against England, all Europe was agog. A Spanish duke went to the Netherlands (where Holland and Belgium are now) prepared to invade England. The Pope blessed the venture. It was easy to predict who would win this fight.

Only it didn't happen that way. When the mighty Spanish armada faced the smaller English navy, some astonishing things happened. The big Spanish galleons weren't prepared for the small, fast English ships. Besides, the Spaniards expected to fight in the traditional way—by boarding enemy ships and fighting hand-to-hand. They were good at that. The English wouldn't oblige. They set old ships on fire and sent them into the crowded armada. (That was Drake's idea.) The winds were fierce, the ships were wooden—and you can guess what happened. The underdog England won. Spain lost 63 ships and about 20,000 men. England did not lose a single ship, except those that were purposely set on fire. Only 100 Englishmen died.

After that, power began to shift. France and England became the nations to watch, especially in North America.

After fireships broke up the Armada, the tattered fleet had to sail northward, around Britain, to get away. Many ships were wrecked.

While the Spanish were preparing the Armada, Drake made a sneak attack. He took 24 ships right into the Spanish harbor of Cadiz, where the warships were at anchor, and captured or destroyed 37 of them. He returned to England and declared he had "singed the king of Spain's beard."

The Spanish Armada met its fate
In fifteen hundred and eighty-eight.

39 The End: Keep Reading

In 1600, life for the First Americans is beginning to change—for good or ill.

This is Chapter 39, and we are at *the end of the first book*, but we're just getting started on the story of US, so *keep reading*. In Book 2 you will find adventure, danger, and some laughs, too.

By now you're good at riding the time capsule. So strap yourself in and set the dial for 1600. We're going to take one more ride over North America.

If you look down you will see swamps, canyons, waterfalls, volcanoes, geysers, plains, mountains, and deserts. Do you see the people? They are hunters, farmers, traders, artists, teachers, and builders, and they have lived on this land for tens of thousands of years. They speak more than 250 different languages. If you look closely, you will see a few newcomers who are fishermen, pirates, explorers, slaves, priests, and soldiers. The newcomers have built tiny settlements in Florida, in New Mexico, and in New France (Canada). Most of them speak Spanish or French. There are other newcomers, too: horses, pigs, sugarcane, and oranges.

If you sniff a bit, you'll catch the the soft fragrance of honeysuckle, pine, and cedar. Listen and you'll hear the music of birds and coyotes. But mostly it is quiet. It is a sleeping giant of a land.

Across the sea, in England, the sounds are loud and insistent: hammers pound, weavers' looms vibrate, and sheep bleat. Do you hear the clamor of people in the cities? Shakespeare's words ring out at the Globe Theatre. This is still the age of Queen Elizabeth—a great age of opportunity for those who are wealthy or lucky. But England has problems it cannot seem to solve. Why are there so many poor people and what is to be done with them? Crime is rampant. Jails are full. The streets are thick with beggars, and many are orphan children.

Soon the Elizabethan Age will come to an end. The great queen has only a few years to live. She has helped make England a world power. English ships now sail confidently around the globe.

A business firm, the London Company, is getting ready to send some Englishmen to America. They are hoping to find gold—or anything else that will make money for them. Those Englishmen will start a colony, and that colony will grow into a nation. It will change the North American continent forever.

It is their story that comes next. But before you get to it, think about what you have learned in this book: about Beringia, about the First Americans, about explorers, about Spaniards, about Frenchmen, about Sir Walter Raleigh and the Lost Colony.

Have you learned something about the ways historians think? Do you realize how much we don't know about the past? How much is yet to be discovered?

You may be surprised at all the people and all the adventures ahead of you before we come to the end of our history.

Actually, you will never get to the end. Books end, but history keeps marching on. That means that someday *you* can write a chapter in the story of US.

This is a page from a book about the New World by an English adventurer, John Smith. Here he is fighting a great Indian chief—but he respects the Indians and their customs, too. In his book he also tells how the Indians worship—the left-hand picture shows one such scene. You will hear more about John Smith in Book 2 of *A History of US.*

Chronology of Events

This is a *chronology* (krun-AHL-uh-jee). It is a list of events in the order in which they happened.

Before you read the chronology, you might be interested to know more about that multisyllable word. First, it is related to every other word that ends in *-ology: archaeology, geology, biology,* and so on. *Ology* means the study of something—or, sometimes, the knowledge gained from that study. *Chronology* is also related to some other words: *chronicle, chronic, chronological.* All these words come from a Greek root—*chronos*—which means "time." And they all have to do with the arrangement of events in time.

A *chronicle* is a record of events; a record that tells a story from its beginning to its end (if it has an end). *A History of US* is a chronicle of United States history. What does *chronic* mean? Something that goes on for a long time, like some illnesses.

If your teacher asked you to put a list in chronological order, what would you do? You'd start back in time, at the earliest date you have,

which is just what we are doing here.

Here we go, back in time, way back—way, way back—through our chronology, to a land inhabited by animals, reptiles, birds, and insects. The story begins in:

circa 38,000 B.C.E. (which was about 40,000 years ago—*circa* is Latin for "about"): a bridge of land was open between Asia and America, and animals went back and forth on it. Some experts say people did too. But no one knows if that is true. If there were travelers, they didn't leave any signs, any records or tools (that we have found). Then the land bridge disappeared. It was covered with water.

circa 26,000 B.C.E.: Beringia (the land bridge) opened up again. Now, did people come across it? No one is sure about that either—there are still no signs or records from our ancient ancestors (although most experts believe they were in America by this time).

circa 10,000 B.C.E.: now we're sure. We have ancient tools and evidence of hunting and camping sites.

What Is B.C.E.?

It means **B**EFORE THE **C**OMMON **E**RA. The Common Era begins with the year 1, which is the year Jesus Christ was (probably) born. Everything before that year 1 is B.C.E. (or B.C., which stands for **B**EFORE **C**HRIST). Everything after the year 1 is called C.E. (or **C**OMMON **E**RA) or A.D. (**A**NNO **D**OMINI, which is Latin for "in the year of

our Lord"). Since Christ is not the Lord in many religions (Islam, Buddhism, and Judaism, for instance), today B.C.E. and C.E. are the preferred initials. Our time is in the Common Era. You can write today's date, put a C.E. after it, and be correct, but after we get past the first 1,000 years C.E., we usually drop those letters.

circa 3,000 B.C.E.: pieces of pottery found in South America tell us that people were making and firing pots on this continent more than 5,000 years ago.

circa 500 B.C.E.: people in Mexico are carving picture writing (hieroglyphs) in stone. They have large settlements, they have irrigation systems for farms, and they work with metals.

circa 300 B.C.E.: the Hopewell civilization is thriving in the Mississippi valley of North America.

Year 1 C.E.

circa 986 C.E.: Viking Bjarni Herjolfsson's voyage

circa 1000: Leif Eriksson settles Vinland

1275–92: Marco Polo travels to Cathay

circa 1300: Anasazi cliff dwellers leave Mesa Verde

circa 1325: rise of the Aztec empire

1400s: formation of Iroquois league

1454: Johannes Gutenberg invents his movable-type printing press

1492: Christopher Columbus's first voyage

1497: John Cabot reaches Newfoundland

1498: Vasco da Gama sails eastward to India

1504: Columbus saved by an eclipse of the moon

1507: Amerigo Vespucci reaches South America

1510: the first enslaved Africans in the Americas

1513: Vasco Nuñez de Balboa sees the Pacific

1513: Juan Ponce de León lands in Florida

1519: Hernando Cortés lands in Mexico

1520: Ferdinand Magellan crosses the Pacific

1520: Disease kills most of Mexico's people

1525: Giovanni da Verrazzano finds New York Harbor and the Hudson River

1532–33: Francisco Pizarro destroys the Inca empire in South America

1534–35: Jacques Cartier sails up the St. Lawrence River and claims the territory for France

1536–39: Estebán and Fray Marcos search for Cíbola in North America's Southwest.

1539: Hernando de Soto's expedition explores the area between Florida and the Mississippi

1540: Francisco Coronado sets out for Cíbola

1542: Juan Rodriguez Cabrillo reaches California

1550: birth of Wahunsonacock, who will found the Powhatan nation

1551: University of Mexico City founded

1562: Jean Ribaut starts a French colony on the coast of what is now South Carolina

1564: French establish Fort Caroline

1565: Spanish destroy Fort Caroline

1568: French destroy San Mateo in Florida

1579: English captain Francis Drake reaches San Francisco Bay, California, on his way around the world

1585: Sir Walter Raleigh's first colony settles at Roanoke Island

1586: Francis Drake destroys St. Augustine and takes surviving Roanoke colonists back to England

1587: John White leads second expedition to Roanoke—the Lost Colony

1588: English fleet defeats the Spanish Armada

1598: Juan de Oñate settles New Mexico

1608: Champlain founds Quebec

1610: Santa Fe founded

More Books to Read

At the end of a history book, most writers make a list of books for further reading. I decided I would tell you about some books that I *love* to read. Mostly, they are historical novels and original documents (which are real words from the past). So here they are, some of my favorite books:

Novels and Other Good Books

Betty Baker, *Walk the World's Rim*, Harper, 1965. The story of Estebán and the search for Cíbola.

Olaf Baker, *Where the Buffaloes Begin*, Viking Penguin, 1985. Little Wolf rides through the night in search of the lake where the buffaloes begin.

Hal Borland, *When the Legends Die*, Lippincott, 1963. A wonderful, exciting, sad story.

Clyde Robert Bullen and Michael Syson, *Conquista!*, Crowell, 1978. A young Indian boy meets his first horse, a refugee from Coronado's expedition.

Tomie de Paola, *The Legend of the Bluebonnet: An Old Tale of Texas*, Putnam, 1983. A Comanche story of She-Who-Is-Alone, a girl who saves her people.

Barbara Esbensen, *The Star Maiden: An Ojibway Tale*, Little, Brown, 1988. What shape should the Star Maiden take when she comes to earth? Read this book and find out.

Jean Fritz, *Where Do You Think You're Going, Christopher Columbus?*, Putnam, 1980. I like all of Jean Fritz's books. So will you. This one is easy to read.

Paul Goble, *Buffalo Woman*, Aladdin, 1984. Can a buffalo become a beautiful woman? It happens in this book.

James Houston, *The White Archer, An Eskimo Legend*, HBJ, 1967. A young Eskimo vows revenge when his parents are killed and his sister captured by a band of Indians.

Gerald McDermott, *Arrow to the Sun: A Pueblo Indian Tale*, Viking Penguin, 1974.

Scott O'Dell, *Sing Down the Moon*, Houghton Mifflin, 1970. Bright Morning has been captured by Spaniards, who wish to make her a slave.

Chester G. Osborne, *The Memory String*, Macmillan, 1984. This story describes how the Indians might

have moved from Sibera to Alaska, 30,000 years ago.

Gary Paulsen, *Dogsong*, Bradbury, 1985. An Eskimo youth goes on a dogsled journey.

Mari Sandoz, *Cheyenne Autumn* (Avon, 1969) and *Horsecatcher* (Westminster, 1973). Sandoz is a great storyteller.

Elizabeth George Speare, *The Sign of the Beaver*, Houghton Mifflin, 1983. You will love this novel about Iroquois life.

Henry Treece, *Westward to Vinland*, S. G. Phillips, 1967. The Vikings come to a new land across the sea.

Jane Yolen, *Encounter*, HBJ, 1992. This is the Columbus story imagined from a Taino Indian's point of view. It is not a cheerful story.

Some of these authors have written many good books. Look for others at your library.

Myths, Legends, and Poems

Joseph Bruchac, *Native American Stories*, Fulcrum, 1991. Fairy tale–like stories of birds, animals, and how the world came to be.

Padraic Colum, *The Children of Odin: The Book of the Norse Myths*, Macmillan, 1948. I love the stories the Vikings told. Colum's book is a classic.

Virginia Hamilton, *In the Beginning: Creation Stories from Around the World*, HBJ, 1988. Indian and other tales of the beginning of the world. A beautiful book.

Henry Wadsworth Longfellow, *Hiawatha*, Dial, 1983. A shortened version of the poem. Reader's Theater Script Service, P.O.Box 178333, San Diego, CA 92117, publishes a play script called *Hiawatha's Childhood*.

Martine J. Reid, *Myths and Legends of the Haida Indians of the Northwest*, Bellerophon, 1988. Just what the title says it is.

Richard Redhawk, *Grandfather's Origin Story: The Navajo Indian Beginning*, Sierra Oaks, 1988. A Navaho creation story.

Virginia Driving Hawk Sneve, *Dancing Teepees: Poems of American Indian Youth*, Holiday House, 1989.

Original Documents

The Log of Christopher Columbus: The First Voyage to America in the Year 1942, Linnet Books, 1989. The Columbus story in his own words (translated from Spanish). This is the log of the first voyage to America—from August 3, 1492, to October 14, 1492. It is printed in big, bold type with good illustrations.

The Log of Christopher Columbus, translated by Robert H. Fuson, International Marine Publishing, 1987. This handsome volume has Columbus's complete log and many interesting notes and illustrations.

Bernal Díaz de Castillo, *Cortez and the Conquest of Mexico by the Spaniards in 1521*, Linnet Books, 1988. This is a real account by a man who was with Cortés. It is illustrated with drawings of the time.

Stefan Lorant, ed. *The New World, the First Pictures of America, made by John White and Jacques Le Moyne*, Duell, Sloan, & Pierce, 1946. The words and pictures in this book come from the 16th century. This is eye-witness writing. If you can find it, check out this book!

Some Nonfiction Books About Early Americans, Vikings, and Explorers

America's Fascinating Indian Heritage, Reader's Digest, 1978. Recommended for reference. Good pictures.

Barbara Brenner, *If You Were There in 1492*, Bradbury, 1991. What was life like in Spain at the time Columbus set sail? This well-written and researched book will tell you. I recommend it highly.

Virginia Pounds Brown and Laurella Owens, *The World of the Southern Indians*, Beechwood Books, 1983. This book tells about the Mound Builders and other Indians of the South. Read about the town of Cofitachequi (koh-fee-tah-CHAY-kee), near today's Augusta, Georgia, and about how de Soto plundered its temple mound, stole 300 pounds of pearls, and took its woman ruler captive.

W. P. Cumming, R. A. Skelton, D. B. Quinn, *The Discovery of North America*, American Heritage Press, 1971. A huge, gorgeous picture book with a scholarly text. The old maps are wonderful.

Roy A. Gallant, *Ancient Indians: The First Americans*, Enslow, 1989. Do you want to know more about mastodons, giant beavers, and Paleo-Indians (Ice Age Americans)? Fascinating facts about them and about those scientific discoverers, archaeologists.

Janet Hubbard-Brown, *The Secret of Roanoke Island: A History Mystery*. Just where did those lost colonists go? Some clues for you to ponder.

Langston Hughes, Milton Meltzer, and C. Eric Lincoln, *A Pictorial History of Black Americans*, Crown, 1983. A story that begins in Africa and continues through black American history.

Philip Kopper, *The Smithsonian Book of North American Indians: Before the Coming of the Europeans*, Smithsonian Books, 1986. An adult reference book with fine illustrations. If you need to do a report about Indians before Columbus's time, this book will be helpful.

Patricia Lauber, *Living with Dinosaurs* (Bradbury, 1991) and *Tales Mummies Tell* (Harper, 1985). These books are fun, full of archaeological information and explanations of techniques such as carbon dating.

Robert Livesey and A. G. Smith, *Discovering Canada: The Vikings*, Stoddart, 1989. Excellent words, games, drawings, and activites—all about the Vikings.

Ann McGovern, *If You Lived with the Sioux Indians*, Four Winds/Scholastic, 1974. Colorful, very easy to read, and informative.

Robin Ridington and Jillian Ridington, *People of the Trail: How the Northern Forest Indians Lived*, Salem House, 1978. Excellent paperback of Woodland Indians and their ways.

John Anthony Scott, *The Story of America: A National Geographic Picture Atlas*, National Geographic, 1984. Written for young people, this book is clear and colorful.

Maia Wojciechowska, *Odyssey of Courage: The Story of Alvar Nuñez Cabeza de Vaca*, Atheneum, 1965.

Charlotte Yue, *The Tipi: A Center of Native American Life*. Knopf, 1984. All about tepees and the way the Great Plains Indians lived.

Index

A

Abenaki Indians, 51
Acoma Indians, 115
Adena Indians, 46, 47
adobe, 30
agriculture, 74, 130. *See also* farming, plants
Alabama, 113
Alarcón, Hernando de, 110
Alaska, 16, 17, 22, 24, 25
Algonquian tribes, 24, 36, 51
  languages, 51
Algonquin Indians, 133, 134, 135
alpaca, 97, 99
Amazon River, 85
Anasazi Indians, 27–30
Anglican Church, 124, 125
animals and birds, 14, 15, 16, 17, 18, 20, 21, 25, 30, 31, 36, 40, 48, 74, 93, 103, 108, 150. *See also* hunting
anthropology, 58, 59
Apache Indians, 27
Apalachee Indians, 103, 112
Appalachian Mountains, 22, 37
Arapaho Indians, 40 ·
Arawak Indians, 69, 70
archaeology, 58
  and mound builders, 44, 88
Arctic Circle, 65, 81
Aristotle, 120
Arizona, 29, 107, 116
Armada, Spanish, 148, 149
armillary sphere, 61
armor, 32, 48, 81, 103, 106, 111, 112
arquebus, 134
arrow. *See* bow and arrow
Asia, 16-19, 22
astrolabe, 67, 68
astronomy, 77
Atahualpa, 99
Atlantic Ocean, 19, 22, 42, 48, 69, 78, 81
atlatl, 19, 20
avocado, 74
Aztec Indians, 87, 88, 91–93, 95

B

Bahama Islands, 69
Baja California, 110
Balboa, Vasco Nuñez de, 78, 79, 110
barbecue (*barbacoa*), 107, 128
baskets, 20, 32, 35
beans, 49, 74, 112

beavers, 17, 132, 133
Beringia, 16–18, 24, 151
Bering Sea, 16, 18, 24
Bering Strait, 16, 18, 19
Bible, the, 61, 76, 118, 120
Bill of Rights, 12
birds. *See* animals and birds
bison, 16, 17
Blackbeard, 131
Black Death. *See* plague
Blackfoot Indians, 40, 51
bones, 14, 15, 16, 58
bow and arrow, 16, 41, 48, 83, 97, 103, 112, 144
Brazil, 101
Bronze Age, 15
buffalo, 36, 41, 42

C

Cabeza de Vaca, Alvar Nuñez, 103, 104
Cabot, John (Giovanni Caboto), 78, 86, 110, 126
Cabrillo, Juan Rodriguez, 109, 110, 111
Cahokia, 46, 47
California, 22, 34, 35, 110, 111, 148
calumet, 113
Canada (New France), 24, 27, 36, 40, 57, 133, 137, 150
Canary Islands, 67
cane, 114
canine, 45
cannibals, 53, 70
cannon, 94, 107
canoe, 20, 31, 34, 35, 96, 126, 136
caravel, 64
Carib Indians, 70
Caribbean Islands, 73
Caribbean Sea, 147
caribou (reindeer), 25
Carlyle, Thomas, 62
Cartier, Jacques, 127
Casas Grandes, California, 34
cassava, 74
Castañeda, Pedro de, 107
Cathay. *See* China
Catholicism, 87, 88, 104, 116, 123, 124, 125, 128
Cavelier, René Robert. *See* La Salle
Cavendish, Thomas, 142
Cayuga Indians, 52
centaur, 91
Cervantes, Miguel de, 12

Ceuola. *See* Cíbola
Chaco Canyon, 29
Challeux, Nicolas Le, 129, 130
Champlain, Lake, 134
Champlain, Samuel de, 133-135, 143
Charles V, Holy Roman Emperor, king of Spain, 117, 120, 121
Charlesfort, 128, 129
Cherokee Indians, 51
Chesapeake Bay, 131
Chesapeake Indians, 37
Cheyenne Indians, 40
Chickasaw Indians, 51
China (Cathay), 19, 63, 64, 66, 67, 69, 70, 72, 74, 75, 80, 85, 86, 126
Chinook Indians, 31
Chippewa Indians, 53
chivalry, 139
chocolate, 74, 93
Choctaw Indians, 51
Christianity, 83, 87, 88, 115, 116, 136
Cíbola, 104-108, 110, 111, 115, 116, 127
Cicero, 11
Clio, muse of history, 11
clothing, 14, 25, 32, 33, 41, 42, 49, 50, 113, 127, 140
cocoa, 93
colonialism, 100, 101
Colorado River, 110
Columbus, Christopher, 19, 37, 60, 63-79, 85-87, 110, 111, 117, 122, 126
Comaco, 79
Comanche Indians, 40
compass, 62, 63, 67, 68
conquistadors, 78, 108, 110, 111, 112, 115, 122
Constitution
  of the Iroquois Confederacy, 54
  of the United States, 12
continents, geography of, 22
Cook, Captain James, 32
core, earth's, 22
corn, 20, 28, 47, 49, 53, 74, 93, 106, 112
Coronado, Francisco Vasquez de, 41, 106-110, 114, 115
Cortés, Hernando, 91-93, 95-97, 103, 104, 110, 115, 127
cotton, 73
creation, stories of, 54, 90
Cree Indians, 40, 51
Creek Indians, 51
crests, 31

Croatan, 145
crops. *See* farming, plants
crossbow, 99, 107
Crow Indians, 40
crusades, 28
crust, earth's, 22
Cuba (Colba), 70, 98
Cuzco, 99

D

dancing, 26, 28, 32, 42
Dare, Eleanor and Virginia, 144
Darien (Panama), 78, 79
dart, 19, 20. *See also* atlatl
Deganwidah, 53, 54
Delaware Indians, 51
democracy, 12
de Soto, Hernando, 109, 111, 112, 114, 117
Díaz del Castillo, Bernal, 96
dinosaurs, 22, 23, 36
disease, 28, 62, 70, 93, 95, 100, 116
dogs, 15, 40, 78, 112
Drake, Sir Francis, 131, 132, 142, 148, 149
drought, 23, 30
Dürer, Albrecht, 98

E

earthquakes, 23
Eastern Hemisphere, 65
Edict of Nantes. *See* Nantes, Edict of
El Dorado, legend of, 97
Elizabeth I, queen of England, 125, 129, 138, 139, 142, 148, 150
Elizabethan Age, 138, 139, 151
Enciso, 79
England, 124, 125, 132, 138-49, 150, 151
Enrique, 83
eohippus (horse's ancestor), 20
equator, 22, 65
Eriksson, Thorvald, 57
Eriksson, Leif, 56, 58
Erik the Red's Saga, 58
Eskimo (see Inuit)
Estebán, 103-106, 110, 116
explorers, 56-7, 63, 66, 72, 74, 77, 78, 79, 80, 85, 91, 97, 106, 110, 111, 150, 151
  English, 138-49
  French, 116-37
  Spanish, 66-123
  Viking, 56-59

F

farming, 20, 29, 30, 40, 42, 49, 52, 56, 73, 130. *See also* plants
Ferdinand, king of Spain, 66, 67, 69, 70, 72, 78, 117, 122, 123
fire, 14, 15, 28
fishing, 17, 18, 20, 24, 31, 53, 70, 114, 128
flat earth, belief in, 60
Florida, 98, 103, 111, 114 127, 129, 130, 150
Folsom, New Mexico, 17
food, 15, 18, 20, 24, 25, 31, 41, 53, 67 77, 83, 107, 112, 115, 128, 133, 139, 144
Ford, Henry, 11
Fort Caroline, 130, 131
fossils, 17, 59
Fountain of Youth, 97
France, 123, 125-37, 143
Francis I, king of France, 117, 126
Franklin, Benjamin, 55
Fray Marcos. *See* Niza, Fray Marcos de)
freethinker, 123
frigate, 147
Frontenac, Louis, 136

G

galleon, 127
game. *See* animals and birds, hunting
games, 20, 29, 32, 35, 51
Ganz, Joachim, 142
Garden of Eden, 75
Garvey, Marcus, 12
Genoa, 64
geography, 64-66, 76
geology, 22, 23
German, 12
Gilbert, Sir Humphrey, 140, 141
glacier, 16, 17, 18, 22, 36
Globe Theatre, 138, 150
gold, 20, 70, 72, 73, 79, 93, 96, 97-99, 102, 103, 106, 108, 109, 110, 111, 122, 123, 127, 144, 148, 151
Golden Hind, 132, 148
government
  Anasazi, 21, 29
  Iroquois, 52, 53, 54
  United States, 8, 12, 13
Grand Canyon, 108, 110
Grand Khan, of China, 67, 70, 71, 74, 109
Great Lakes, 37
Great Turtle, 54, 90
Greenland, 24, 56

Greenlanders, The Saga of the, 59
griffin, 110, 111
Guam, 82-84
Gudrid, 57, 58
Gulf of California, 110
Gulf of Mexico, 137
Gulf Stream, 127
guns, 98, 99, 107, 111, 113, 134, 144
Gutenberg, Johannes, 61-63

H

Haida Indians, 31, 33
Hakluyt, Richard, 105, 133
hammock, 20, 71
Harriot, Thomas, 142
Hatuey, 123
Haudenosaunee, 51. *See also* Iroquois Indians
Hawikuh, 107
Henry, Prince, the Navigator, 63, 66
Henry IV, king of France, 128
Henry VIII, king of England, 117, 124
Herjolfsson, Bjarni, 56
Hiawatha, 53, 54
Himalayan Mountains, 22
Hispaniola, 71, 72, 74, 78, 122
Homo sapiens, 14
Hopewell (ship), 147
Hopewell Indians, 45, 47, 58
Hopewell mounds, 44, 45
horizon, 145
horse, 20, 21, 42, 91, 103, 106, 107, 108, 111, 112, 150
housing, 14, 16, 25, 26, 27, 28, 30, 32, 41, 51, 58
Huguenots (French Protestants), 125, 127-129, 135
hunting, 14, 16, 17, 19, 20, 21, 24, 25, 30, 32, 40, 41, 42, 48, 50, 52, 105. *See also* animals and birds
Huron Indians, 48, 51, 53, 133, 134, 135

I

Ice Age, 14, 16, 17, 20
igluviga (igloos), 25, 26
Illinois, 136
Illinois Indians, 36
Illinois River, 46
Inca empire, 97–99, 101
India, 22, 83
Indians, mound building, 43, 44, 47, 49
  of Northwest, 31, 32

of Plains, 40–42
of Southwest, 27, 30. *See also* Native
  Americans *or under name of tribe*
Indies, 19, 63, 69
Inquisition, Spanish, 89, 122, 123
Inuit, 24–26
Iowa Indians, 40
Iron Age, 15
Iroquois Indians, 48, 51–53, 88, 90
Iroquois League, 52, 54, 55
Isabella, queen of Spain, 66, 70, 72,
  78, 87, 89, 117, 122, 123
Isthmus of Panama, 148
Italy, 78, 126
Iztapalapa, 96

J

Jackson, Jesse, 9
jackstaff, 63
Jamaica, 77
javelin, 83
Jefferson, Thomas, 43, 55
Jesuits, 136
Jews, expulsion from Spain of, 88,
  122, 123
John, king of Portugal, 75
Joliet, Louis, 136
Judaism, 87, 88, 122, 123

K

Kalapuya Indians, 31
Kalispel Indians, 31
Kansa Indians, 40
Karlsefni, Thorfin, 57, 58
kayak, 26
Kennedy, John F., 11
Kensington Stone, 60
Kino, Francisco Eusebio, 116
Kipling, Rudyard, 90
kiva, 28, 29
Koran, 120
Kwakiutl Indians, 31, 32

L

lacrosse, 20
lance, 111, 112, 113
language, 20, 24, 34, 50, 51, 52, 67,
  70, 101, 112, 122, 150
Lapu Lapu, 83
La Salle (René Robert Cavelier), 137
Las Casas, Bartolomé de, 117–22
latitude, 64–66, 68

lava, 23
League of the Great Peace, 54
LeMoyne, Jacques, 21, 87, 105, 130
Leoncico, the dog, 78
llama, 17
London Company, 151
Longfellow, Henry Wadsworth, 53
longhouse, 52, 55
longitude, 64–66, 68
Lopez de Cardenas, García, 108
Lost Colony. *See* Roanoke Island
Louis XIV, king of France, 128–37
Louisiana, 137
Luther, Martin, 87, 88, 122

M

Madoc, prince, of Wales, 58
Magellan, Ferdinand, 81–84, 149
magma, 22, 23
magnetic north, 68
Mahican Indians, 51
maize, 106. *See also* corn
Malinche, La, 93
mammoth, 15, 16, 20, 40
Mandan Indians, 40, 42
mantle, earth's, 22
mapmaker, 63, 64, 85
maps, 64, 65, 66, 86
Marina, Doña, 92, 93
Marquette, Jacques, 136
Mary I, queen of England, 124, 125, 138
Massachuset Indians, 51
Maya Indians, 90
Menendez de Aviles, Pedro, 130,
  131, 138
meridians, 66. *See also* longitude
Mesa Verde, Colorado, 27, 28, 30
mestizos, 101
Mexico, 21, 87, 100–104, 106, 110,
  122, 123
Mexico City, Mexico, 95
missionary, 116
mission churches, 116
Mississippi River, 36, 37, 40, 41, 43,
  46, 48, 114, 136, 137
Missouri River, 36, 46
Moctezuma (Montezuma), 91–95,
  97, 115
Mohawk Indians, 52
Moluccas. *See* Spice Islands
Mongolia, 14
More, Sir Thomas, 140, 141

mound builders, 43, 44, 46, 47, 49
mounds, 43, 44, 47, 49
mulattoes, 101
mule, 21, 115
musical instruments, 15, 29, 32, 103, 113
musket, 144. *See also* guns
musk ox, 18
Muslim, 87, 88, 120, 122

N

Nantes, Edict of, 128
Narvaez, Panfilo de, 103, 112
Native Americans, 19, 20, 38, 39,
  118, 120. *See also* Indians *or under
  name of tribe*
Navaho Indians, 27
navigation, 62, 63, 67, 68
Neanderthal, 14
Netherlands, 139, 149
Newfoundland, 58, 78, 126, 133
New France. *See* Canada
New Mexico, 114, 115, 116, 150
New Spain. *See* Mexico
Niña, 64, 67
Niza, Fray Marcos de, 103, 105–107,
  110, 116
Nootka Indians, 31, 32, 33
Norsemen. *See* Vikings
North Pole, 24, 65
Northwest Passage, 126, 136
Nova Scotia, 56, 59

O

oceans, 18, 57, 68. *See also*
  Atlantic Ocean, Bering Sea, Pacific
  Ocean, Sargasso Sea
Ohio, 43, 47
Ohio River, 48
Ojibwa Indians, 36, 40, 51
Olmos, Alonzo de, 131
Omaha Indians, 40
Oñate, Juan de, 115–117
Oneida Indians, 52
Onondaga Indians, 51, 52
Outer Banks, 145, 147

P

Pacific Ocean, 66, 80, 82, 148
Palos, Spain, 67
Pangea, 22
parallels, 65. *See also* latitude
passenger pigeon, 48

Patagonia, 81, 138
peppers, 20, 74
Peru, 79, 98, 99, 101, 123, 148
Philip II of Spain, 124
Philippine Islands, 83
Pigafetta, Antonio, 82-84
pineapple, 72, 74
Pinta, 64, 67, 69
pirates, 126, 127, 130, 131
Pizarro, Francisco, 79, 98, 99, 106, 110, 111, 117
plague, 100
Plains Indians, 40–42
plants, 20, 35, 37, 73, 74, 108, 112, 113, 114, 150. *See also* farming
Pole Star, 63
Polo, Marco, 63, 64, 69, 74, 109
Pomeiock, Virginia, 146
Ponce de León, Juan, 97, 98, 110
Popocateptl, 93
Popol Vuh, 90
Portugal, 63, 64, 66, 75, 83, 101, 124, 143
potato, 20, 74, 106
potlatches, 33
Potosí, Peru, 101
pottery, 20, 28, 58
printing, 59, 61, 62, 85, 99, 101
privateer, 126, 127, 130, 148
Protestants, 87, 88, 122, 123, 124, 125, 128
Ptolemy, 65, 66, 81
pueblo, 29, 30, 102, 107, 116
Puerto Rico, 97
pumpkin, 74, 112

Q
Quebec, 134, 135, 143
Quetzalcoatl, 91, 92, 93, 94
Quinault Indians, 31
Quivira, 109

R
radio carbon dating, 58
Raleigh, Sir Walter, 138, 140-143, 151
religion, 87–90, 99, 101
Renaissance, 62
Ribaut, Jean, 127-129, 131, 138
rifle, 42
Rio de Janeiro, 85
Rio Grande, 30, 115
Roanoke Island, 144, 145, 147, 151
Rocky Mountains, 23, 36, 40

S
sachem, 52
sailing, 17, 56, 58, 63, 64, 68, 99, 126, 127, 128, 151
St. Augustine, Florida, 130, 134, 143
Salem witch trials, 60
San Diego Bay, 110
San Mateo, 131
San Salvador, 69, 79
Santa Elena, 131
Santa Fe, 115, 116, 134, 143
Santa María, 64, 67, 75
Santangel, Luis de, 71
Sargasso Sea, 68
Scandinavia, 56
Schmidel von Straubing, Ulrich, 103
Seneca Indians, 52
Sepulveda, Juan Ginés de, 120, 121
Serpent Mound, 43, 44, 47
Shakespeare, William, 138, 150
ships, 57, 58, 64, 67, 75, 94, 126, 127, 132, 147, 149
Sierra Nevada, 35
silver, 97, 95, 99, 101, 109, 115, 122, 124
Sioux Indians, 40
slash-and-burn farming, 49
slavery, 12, 32, 46, 70, 72, 73, 74, 88, 112, 117–21
smallpox, 70, 94, 95, 100
Smith, John, 151
Snorri, 58
snowshoes, 20, 24, 26
South America, 17, 18, 36, 75, 80, 85, 98, 101, 122
Southwest Indians, 27, 30
Spain, 21, 35, 66, 70, 74, 78, 80, 91, 97, 101, 102, 106, 110, 116, 117, 120, 121, 122, 123, 124, 127, 130, 131, 132, 139, 143, 148, 149
Spice Islands (Moluccas), 80, 82, 83
steam huts, 35
stinkards, 46
Strait of Magellan, 80, 81, 148
succotash, 49
sugar (cane), 73, 74, 101, 121
sunflower, 74, 146

T
Tadodaho, 53, 54
taiga, 26
Taino Indians, 69, 70
Tenochtitlán, 92–96

tepee, 41
Texas, 40, 114, 137
Texcoco, Lake, 96
Tierra del Fuego, 82
Tlingit Indians, 31, 32
tobacco, 70, 101, 121, 141
Toltec Indians, 93. *See also* Aztec Indians
tools, 15, 31, 40
totem poles, 31
tribe, 34
tribes, Indian. *See* Indians, Native Americans, *or name of tribe*
tundra, 25, 26
"Turk, the," 108, 109
Turtle Island, 54, 55
Tuscarora Indians, 52

U
umiak, 26
Utopia (Sir Thomas More), 141

V
Venezuela, 119
Verrazano, Giovanni da, 126
Vespucci, Amerigo, 85, 86, 110
Vikings (Norsemen), 12, 56, 57, 60, 78
Vinland, 56, 59, 60
Virginia, 132, 142
Vives, Juan, 13
volcanoes, 23

W
Waldseemüller, Martin, 86
wampum, 52, 54
weapons, 112, 134, 144. *See also* bow and arrow, crossbow, guns, javelin, lance, musket, rifle
West Indies, 19, 63, 69, 74, 104
wheels, 21, 115
White, John, 50, 144, 145, 147
Wichita Indians, 40, 41
wigwam, 51
Woodlands Indians, 42, 49
Wyandot language, 48
Wyoming Basin, 36

Z
Zacuto, Abraham, 77
Zheng He, 75
zoology, 59
Zuñi Indians, 27, 30, 105, 107

A Note from the Author

My editor says that I will have to wait until Book 10 of A History of US to write a proper "thank-you" page. So I can't tell you about the boys and girls who read the manuscript and told me what they thought of it. Or about the teachers and educators and friends who encouraged me. Or about the people at the American Federation of Teachers and the Smithsonian Institution who believed in this project. Or about the foundations and individuals who helped. But I couldn't have written this book without them.

About Me

Some children's books are written by committees, but not this book. It was written by a real person—me, Joy Hakim. And, since that seems to be unusual for a history series, my publisher asked me to tell you something about myself. Here goes:

When I was six months old my parents entered me in a beauty contest and I WON! I like to make note of that, because it never happened again.

I was born in Forest Hills, New York. But my family soon moved to Rutland, Vermont, where I grew up (along with my brother, Roger). I hoped to be a skiing instructor, but since I get scared when I'm going fast, I had to change career plans. So I went off to college—Smith College—which turned out to be a good choice. I liked the place and learned some things, too.

Then I headed for New York to work as an assistant editor for a foreign news service. That means that I took stories from reporters all over the world and sent them to editors who could use them. I loved that job, but I decided that I wanted to teach. I needed a degree in education to do that and got one at Goucher College. I taught: third grade in Baltimore, Maryland; seventh grade in Syracuse, New York; fifth grade in Omaha, Nebraska; and, in Virginia Beach, Virginia, high school English and college

While this was going on I got married, to Sam, and before long we had three children: Ellen, Jeffrey, and Daniel—which was the best thing I ever did.

Writing and teaching are related fields—they both have to do with explaining and learning. So don't be surprised at the next job I got. It was as a newspaper reporter.

As a journalist I interviewed people. I wrote about schools. I wrote about businesses. I wrote about books. I wrote editorials. I was having a good time doing that kind of thing when Ellen took a history course, and that got me thinking. You see, the teacher taught history backwards.

Yes, you read that right. Backwards! Now why would he do that? When I saw the history book I understood why. The book was so dull that it didn't matter whether you read it backwards or forwards. I decided I could do better. I decided I would spend a year and write a history book for children.

Which just shows how ignorant I was, because it was more than seven years before I finished writing *A History of US*, and the one book turned into ten books.

And that is the story of ME.